ENVIRONMENTAL EFFECTS OF UTILISING MORE COAL

Special Publication No 37

Environmental Effects of Utilising More Coal

The Proceedings of a Conference organised by
The Council for Environmental Science and
Engineering

Royal Geographical Society, London, 11th and 12th
December 1979

Edited by
F. A. Robinson,
Chairman, CESE

Royal Society of Chemistry
Burlington House, London, W1V 0BN

Copyright © 1980
The Royal Society of Chemistry

All Rights Reserved
No part of this book may be reproduced of transmitted in any form or by any means—graphic, electronic, including photocopying, recording, taping or information storage and retrieval systems—without written permission from the Royal Society of Chemistry

British Library Cataloguing in Publication Data
Environmental effects of utilising more coal.—
   (Royal Society of Chemistry. Special publications; no. 37
   ISSN 0577-618X).
   1. Coal—Environmental aspects—Congresses
   I. Robinson, Frank Arnold
   II. Series
   662'.62   TD195.C58
   ISBN 0-85186-805-3

Thanks are due to Renate Majumdar for re-typing the papers presented here.

662.62
ENV

Printed in Great Britain by
Whitstable Litho Ltd., Whitstable, Kent

Preface

The Council for Environmental Science and Engineering (CESE) was set up jointly by the Council of Engineering Institutions (CEI), combining 16 professional engineering bodies, and the Council of Science and Technology Institutes (CSTI) comprising 5 of the professional science bodies, to provide an organisation for keeping under review environmental problems that may arise, particularly as a result of new technology, and for seeking the adoption of realistic solutions to such problems.

The Council has for many years considered the implications of the declining output and increasing cost of oil, and has studied the various options that might be available, e.g. increased use of coal and nuclear energy, possible use of tidal, wave and solar energy and other 'renewable' forms of energy. Each of these presents problems, many of which can be described as environmental, which must be examined by professionally competent people before their merits can be properly assessed. The problems of nuclear energy have had worldwide publicity and have often aroused deep emotions and serious controversy. On the other hand the increased use of coal, of which this country has considerable reserves, has not aroused so much interest.

Coal and nuclear energy are currently the only sources of energy in this country that can replace oil within the near future. The consequences of utilising the 170 million tons of coal per annum envisaged in the National Coal Board's 'Plan for Coal' have not yet been fully assessed.

Whilst other bodies concerned with the environment have devoted most of their attention to problems of nuclear energy, the CESE has thought it of more immediate importance to consider the problems that will arise from the expansion of coal production. If coal is to be utilised in the way it has been in the past, increasing amounts of carbon dioxide, sulphur dioxide and nitrogen oxides will be generated. There may be greater emissions of dark smoke, particulate matter and carcinogens into the atmosphere, and the conversion of coal into gas, liquid hydrocarbons and solid fuel by the old gasworks process could produce serious local pollution. Fortunately, new and improved technologies have been developed that will greatly reduce the serious pollution that might otherwise occur.

The CESE decided that an authoritative review of the consequences of increased amounts of coal being burnt or converted into other products would be of considerable interest to engineers, scientists and all those concerned about the possible impact on industry of an impending shortage of fuel or liquid hydrocarbons as raw material for the production of chemicals. Most of the potential environmental hazards that might result from using coal in place of oil (and in due course in place of gas), for generating electricity, for providing domestic heating and industrial power, and for producing liquid hydrocarbons for chemical production were discussed at the conference. The papers presented and the resulting discussion are collected together in this publication.

Contents

Session I

Chairman's Address of Welcome *F.A. Robinson*	1
Keynote Speech *G. England*	7
Carcinogens from Coal and Other Sources *P.J. Lawther*	14
The Meteorological Effect of Increasing the Carbon Dioxide Content of the Atmosphere *Sir John Mason*	19
Discussion on Session I	41

Session II

Chairman's Opening Remarks *R. Press*	44
Emission of Sulphur Dioxide, Nitrogen Oxides, and Particulate Matter from Coal-Burning Power Stations *A.J. Clarke*	48
The Effects of Changing Emissions on Biological Targets *M.W. Holdgate*	71
Legislative Matters Relating to the Burning of Coal *D.H. Napier*	90
Discussion on Session II	103

Session III

Chairman's Opening Remarks *G.S. Hislop*	107

New Technology in Coal Combustion *T.H. Kindersley and P.C. Finlayson*	111
Environmental Effects of Producing Substitute Natural Gas *F.E. Dean and B. Goalby*	127
Production of Liquid Fuels from Coal *J. Gibson and D.W. Gill*	144
Discussion on Session III and Chairman's Closing Remarks	161

Session IV

Chairman's Opening Remarks *F.A. Robinson*	168
Some Environmental Aspects of the Production of Solid Smokeless Fuel *D.J. Davison*	169
Discussion on Session IV and General Discussion	190

Summing-up and Closing Address

Introduction *F.A. Robinson*	195
Summing-up and Closing Address *Sir Hermann Bondi*	196
Closing Remarks	204

Session I

Chairman's Address of Welcome

F. A. Robinson
CHAIRMAN, COUNCIL FOR ENVIRONMENTAL SCIENCE AND ENGINEERING

I have very great pleasure in welcoming you on behalf of the Council for Environmental Science and Engineering to this Conference which is only the second we have held; you could therefore be forgiven for claiming that you had never heard of us before! The Council was set up a few years ago jointly by the Council of Engineering Institutions, comprising 16 professional engineering bodies, and the Council of Science and Technology Institutes, comprising 6 of the professional science bodies. The objective was to study environmental problems, particularly those of a long-term nature that may simultaneously affect several of the science and engineering bodies, making it desirable to have joint consultations before recommending the adoption of a particular attitude or line of action. It is relatively easy - I did say relatively - to determine what action should be taken after an actual disaster. For example, if damage to the environment had resulted from faults in an aeroplane or its engine, the responsibility in that event is clearly that of the aeronautical engineer though he may of course, according to circumstances, consult his colleagues in the Institution of Electrical Engineers or the Institution of Mechanical Engineers. Similarly an accident in the chemical industry such as that at Seveso will concern the chemical engineer in the first place but is very likely to involve the chemist or the physicist or the metallurgist in due course.

However, the thinking behind the setting up of CESE was based on a rather different philosophy. This was that a group should be formed, each member having the expertise of his own special discipline acquired by practical experience, to pool their knowledge and consider and discuss together the environmental problems that might arise in the future as changes take place resulting from the adoption of new technology. Thus, our earlier conference on the 'Transport of Hazardous Materials' involved the collaboration of many of our member-institutions, since it necessitated amongst other things a review of the wide variety of chemicals that are nowadays being routinely and regularly carried on our roads in Britain and the wide variety of hazards that these would cause in the event of a spillage or an accident, the design of road vehicles and tank containers, the routing of vehicles, and the systems of traffic control and police action following an accident. The coming together of groups of people who in the ordinary course of events would never meet, and perhaps hardly know of one another's existence, proved very valuable in disseminating ideas more widely and promoting discussion between groups with different types of expertise. It may also have helped some of the public to realise that the technical people involved in environmental hazards are more aware of the problems than most people, and are actually engaged in trying to find satisfactory solutions. The substantial sales of the printed proceedings are an indication of the degree of interest created.

The idea of the present conference goes back for many months and is of course based on the recognition that supplies of oil have a finite limit, and will one day come to an end. Oil and natural gas overtook coal as the major source of energy in most parts of the world in 1955. By 1970 the total world demand for energy had doubled, but during those intervening 15 years the consumption of oil and gas had actually increased three-fold, and this trend has continued throughout the last decade. In the United Kingdom, the use of coal actually declined after 1955 whereas oil imports continued to increase.

It was around 1970 that the warning bells began to ring and it was suggested that the world's oil resources might begin to reach exhaustion in about 30 years time. Yet during that time the world's population will increase from around 4 billions to 7 billions with a resulting increase in the demand for food, shelter, fuel, and industrial production and the increased standard of living that the West has long enjoyed in comparison or rather in contrast to the Third World. Whereas in 1955 the non-Communist world consumed 2350 mtce of energy in all forms, it has been estimated that by the year 2000 it will be consuming 20,000 mtce of energy - nearly ten times as much. What is certain is that this will not be coming from oil by that time, for, added to the more widespread use of oil as a fuel and as a chemical feedstock in recent years we have had first successive increases in the price of oil starting in 1973 and continuing almost annually ever since, and then the Islamic Revolution of 1977 as a result of which Iran has almost ceased to be an exporter of oil. Where then will our energy be coming from in the year 2000, and what environmental damage will result from the exploitation of the new sources of energy unless precautions are taken now? The Institute of Petroleum recently held a conference at which the environmental damage caused by oil to this country and the seas surrounding it was reviewed. The CESE believes that now is not too soon to begin assessing the type of damage that might result from these new sources of energy, before they begin to be used on the scale that will be necessary in the future. Moreover a group of professional engineers and scientists are not likely to rest content just with defining the problem, but will want to suggest ways of achieving the ends desired without causing the environmental damage that at the moment may seem to be inevitable, or at all events will admit that there are risks but will indicate how these risks can be minimised, monitored, and controlled. They are responsible and competent people who have the skill and ability to carry out innovatory projects in the public interest.

On paper the world has a number of options open to it as oil becomes scarcer or more expensive. Many of these are hailed by some people as cheap and readily available sources that will last for ever - the sun and the winds, the tides and the waves and 'hot rocks'. Yet the only sources that the CESE believes can be made available within the next decade or two - which is the critical period when they are going to be needed - are coal and nuclear energy. Discussion of the latter arouses strong emotions, although we in this country have been feeding energy from this source into the National Grid for the past 20 years without serious incident. But the problem of producing sufficient energy from coal, and a target of 170 mta was set by the National Coal Board in 1974, seemed to us to be of more immediate urgency. Of course when we came to consider the idea in detail, it proved to be more complicated than we had at first thought, and it was suggested to us that a conference covering the whole gamut of environmental hazards associated with coal production would be too big an undertaking. We therefore agreed to leave out of consideration the problems of getting coal out of the ground and concentrate our efforts on the utilisation of coal, which is complicated enough when one thinks of the many different ways of using it other than simply burning it - e.g. producing liquid hydrocarbons for cars and planes, gaseous and liquid hydrocarbons for domestic heating and chemical feedstock, and solid smokeless fuel and coke. Perhaps the public does not realise what social changes would be forced upon us if scientists and engineers were unable to solve these problems. In any event it is most unlikely that present prices will be maintained. If for example the price of motor fuel rises considerably, as I think it will, shall we be able to continue to enjoy private transport, and will public transport be adequate for all who will then want to use it? Will the cost of basic chemicals have to be increased so much that the cost of plastics, pesticides, paints, drugs, and other consumer goods will have to be considerably increased? Above all, of course, as the supply of oil runs out will the cost of basic fuel supplies have to be increased so much as to affect substantially the price of steel, cement, textiles, foodstuffs, and manufactured goods in general?

We have here an audience - not so large perhaps as we had hoped - drawn from a wide variety of backgrounds, from the coal and electricity generating industries, the oil and gas and nuclear power industries, the Departments of Energy, Environment, and Industry, and from local authorities and the universities. There will be ample time for discussion and we hope you will feel free to question and if you like disagree with any of our speakers. May I end by two brief comments on the papers. Sir John Mason has amended the title of his talk to 'The <u>Meteorological</u> Effects of Increasing the Carbon Dioxide Content of the Atmosphere', and Sir Hermann Bondi in his 'Summing up' will not summarise what has been said by those who have spoken before him, but will sum up the situation as he sees it nationally and I hope also internationally.

Mr. G. England, BSc(Eng), CEng, FIEE, FIMechE, FBIM
Chairman of the Central Electricity Generating Board

After working in applied research, Mr England read electrical and mechanical engineering at Queen Mary College, London University and served with the Royal Electrical and Mechanical Engineers from 1942 until 1947 when he joined the Central Electricity Board.

On nationalisation, in 1948, Mr England served in the London Division of the British Electricity Authority as a technical engineer and, from 1951, in the Generation Design Branch at the Central Electricity Authority headquarters. From 1958 he was responsible, as Development Engineer (Policy) with the Central Electricity Generating Board, for building up a multi-disciplinary planning team and became Chief Operations Engineer in 1966. He was appointed Director-General of the CEGB's South Western Region in 1971, and in August 1973 became Chairman of the South Western Electricity Board. Mr England was appointed a part-time Member of the CEGB in July 1975, and succeeded Sir Arthur Hawkins as Chairman in May 1977.

Professor P.J. Lawther, FRCP, MRC
Toxicology Unit, St Bartholomew's Medical College

P.J. Lawther is Professor of Preventive and Environmental Medicine in the University of London at the Medical Colleges of St Bartholomew's and London Hospitals. He is Physician-in-Charge of the Department of Environmental and Preventive Medicine at St Bartholomew's Hospital and Head of the Clinical Section of the MRC Toxicology Unit. Before reading medicine he worked in the chemical industry. He and his colleagues have made special studies over the last twenty-five years of the nature and effects of air pollutants on man.

Sir John Mason, CB, DSc, FInstP, FRS
Director-General, Meteorological Office

Sir John Mason graduated BSc (1947) at the University of London and was awarded DSc in 1956. He was appointed Director-General of the Meteorological Office in 1965. Previous appointments include Warren Research Fellow of the Royal Society (1957), Visiting Professor of Meteorology, University of California (1959-60), Professor of Cloud Physics at Imperial College London (1961-65).

He was elected a Fellow of the Royal Society in 1965, was President of the Royal Meteorological Society 1969-70, Honorary General Secretary of the British Association for the Advancement of Science (1965-70), is Permanent Representative of the United Kingdom at the World Meteorological Organization and a Member of its Executive Committee. He was President of the Institute of Physics from 1976-78 and is currently Treasurer and First Vice-President of the Royal Society.

He was awarded the Charles Chree Medal and Prize of the Physical Society (1965), the Glazebrook Medal of the Institute of Physics (1974), the Rumford Medal of the Royal Society (1972) for his discoveries on clouds, rain and especially the origin of lightning on which he delivered the Bakerian Lecture of the Royal Society in 1971. He was awarded the Symons Memorial Gold Medal of the Royal Meteorological Society (1975). He was elected Pro-Chancellor of the University of Surrey with effect from 1 September 1979.

Keynote Speech

G. England
CHAIRMAN, CENTRAL ELECTRICITY GENERATING BOARD, LONDON

Thank you for inviting me to give the keynote speech at this conference. The subject is both topical and important. Only two proven sources of energy are at present available to fill the gap which will arise as oil and gas reserves become depleted: one is nuclear energy and the other is coal. But although the effects of increasing use of nuclear energy are frequently debated, far less public attention is being given to the environmental effects of utilising more coal.

This is perhaps understandable. Many people still regard nuclear power as new and mysterious, and therefore suspect, even though it is well established as a means of generating electricity. By contrast coal is regarded as uncomplicated and familiar. We have been living more or less happily and comfortably with it for many centuries, and it has become almost hallowed by tradition. But as papers to this conference demonstrate, strict safeguards are just as essential for burning coal, or any other fossil fuel, as for using neclear fuel.

This truth deserves to be more widely recognised in the national energy debate. Your Council is therefore to be congratulated on its initiative in convening this conference, and thus encouraging an authoritative review of the possible consequences of burning more coal or using it to replace liquid hydrocarbons as a raw material for the production of chemicals.

For the benefit of those of you who are not closely acquainted with the Generating Board's activities, perhaps I should present my credentials.

Electricity production in England and Wales, the area covered by the CEGB, was built up on coal. For many years, indeed, it was the electricity supply industry's only fuel for raising steam. We are today a very large consumer of coal. In fact, we are burning more coal than ever before - between 75 and 80 million tonnes a year - and we expect to maintain that level of consumption for some years to come.

Consistent with the Acts of Parliament which established the CEGB, we have two concerns: to provide economical and reliable supplies of electricity and to do so in a way that takes account of the effects of our activities on the environment.

These concerns have led us to make many detailed studies of the combustion process and of the dispersion of flue gases into the atmosphere. Our claim to fame is that we began our studies long before care for the environment became as fashionable as it is today.

Let me now indicate the aspects of the conference theme which I intend to mention. I shall say a little about the nation's history of concern over the environmental effects of burning coal. I shall move on to examine what lies behind Britain's success so far in reducing air pollution, and consider (with the aid of a cautionary tale or two) how our existing approach can continue to serve us well. I shall also refer to some of the wider implications of burning more coal. I may even mention, from time to time, the work of the Generating Board.

The electricity supply industry's use of coal has a long history, but general concern in Britain about the environmental impact of coal-burning stretches back very much further.

In 1285 Roger de Northwode, John de Cobbeham, and Henry de Galleys were commissioned 'to enquire touching certain Lime Kilns constructed in the City and Suburbs of London and Southwerk of which it is Complained that where as formerly the Lime used to be Burnt with Wood it is now Burnt with Sea Coal, whereby the Air is Infected and Corrupted to the Peril of those Frequenting and Dwelling in those Parts'. In executing this commission they were to associate themselves with 'the Mayor and Sherrifs of London and the Baliffs of Southwerk'.

What, if anything, resulted directly from this inquiry I do not know. But what is known is that in 1307 a Royal proclamation was issued banning the industrial use of sea-coal, for environmental reasons, in London and Southwark. Brushwood and charcoal were the recommended substitutes. Penalties were laid down for using sea-coal - 'from which', said the proclamation, 'is emitted so powerful and unbearable a stench'.

Again, the diarist John Evelyn in 1661, in his pamphlet 'Fumifugium', protested against the 'horrid Smoake' of London, 'which,' he wrote, 'obscures our Churches, and makes our Palaces look old, which fouls our Clothes, and corrupts the Waters, so as the very Rain, and the refreshing Dews which fall in the several Seasons, precipitate this impure vapour which spots and contaminates whatever is exposed to it'.

Of course, we do not have to go anything like as far back as that for some formidable examples of what can happen from the burning of coal. It is, after all, not so long ago that London was notorious for its smog.

It is, however, a heartening augury for the future that the United Kingdom has perhaps been more successful than any other industrialised country in controlling pollution - especially air pollution. In recent decades we have experienced a major improvement in air quality, both gaseous and particulate. In fact, our urban air has become progressively cleaner over a period when fossil fuel consumption in power stations has risen significantly. This is the very reverse of the doomsday prediction that all increases in fuel consumption would increase pollution. Over the past 25 years the consumption of coal in power stations has doubled, while the concentration of sulphur dioxide in the air in towns has more than halved.

This dramatic reduction in pollution is partly the result of wise legislation. This, while forming the basis for firm pressure for improvement from the regulatory authorities, is nevertheless flexible enough to take advantage of steadily increasing knowledge. Its beginning can be traced back more than a century - to the Alkali Act of 1863. This act was mainly concerned with protecting the public from the effects of the manufacture of hydrochloric acid and other chemicals, but it was the first significant move to control, in the modern sense, the air pollution caused by large industries.

The marked reduction in pollution in recent times stems also from the movement of combustion from small open fires in homes and offices to large, centralised power station boilers. This may well have made the greatest single contribution towards reducing the air-pollution effects of coal-burning. It has been estimated that a single tonne of coal burned in a traditional domestic open fire has as much adverse effect on air quality as 25,000 tonnes of coal burned in a modern power station boiler.

Research into the environmental effects of coal combustion can also be traced back over many years. A personal experience is relevant here. Almost exactly 21 years ago - on 4 December 1958 - I led a CEGB investigation called 'Operation Chimney Plumes'. This involved taking aerial photographs of power station plumes rising through low-level fog over London. The investigation proved conclusively that, at the times when low-level sources of pollution were contributing significantly to London's bad air, our power stations were making little or no addition. The plumes, having risen above the belt of fog, dispersed in the clear air above without affecting London air at ground level.

The research work has underlined the importance of sound data and of care in their interpretation. This may be further illustrated by two topical examples. The first shows what can result from a lack of research.

The USA has pursued a policy of imposing flue-gas desulphurisation on fossil-fired power stations. One of the claimed justifications for this was the adverse effects on health of sulphate particulates in the atmosphere. However, the results of the study known as CHESS (Community Health Environment Surveillance System), which were intended to supply this justification, were shown by a Congressional Committee to be of little value as a scientific basis for air-quality standards. Moreover the US Environmental Protection Agency has recently repeated and extended limited early work with sulphates on animals. These results, too, do not support the earlier suggestions. Indeed the most common form of sulphate (ammonium sulphate) produces no detectable toxological effects in rats at concentrations some hundreds of times those to be expected in polluted air. Proper research at the right time would probably have avoided unnecessary alarm and unnecessary restrictions.

Now an example of timely research. At Stockholm in 1972 the Scandinavian countries projected a picture of massive long-distance transport of sulphur dioxide from the United Kingdom transforming into acid rain which harmed their forests and flowed into their lakes and rivers, with disastrous effect on their fish populations. You will hear during the course of this conference how current research is demonstrating that the problem is by no means as simple as it first appeared. The chemistry of rain is complicated; the long-term trends in its composition are subject to considerable uncertainty, and its contribution to surface-water acidity is still under investigation. At this stage one cannot predict the change in acidity that would result from a given reduction in sulphur emissions, but whatever the answer it is already clear that the cost/benefit ratio of controlling sulphur at source would be very high. If the research shows that this approach is of little value, we may well succeed in avoiding restrictive legislation and heavy (and unnecessary) expenditure.

That research is of interest to a number of countries. Of interest to the whole world is the research, which involves many disciplines and is likely to be mentioned more than once at this conference, into the probable consequences of the release of additional carbon dioxide to the atmosphere.

Taking costs and benefits rather more widely, there is a need to maintain a sensible balance in the use of resources. For clean air measures may conflict with economy in energy use.

This will be discussed during this conference and it will be enough for me to remind you of the graphic observation made in 1978 by Professor Sir William Hawthorne, in his Trueman Wood Lecture to the Royal Society of Arts. He pointed out that the extra fuel needed to reach and maintain all the emission standards currently promulgated in North America exceeds the commercial fuel consumption of India. He went on to say: 'It is important to protect our environment but over-protection can use energy which may be needed in the future to keep people warm and employed'.

I have dealt at some length with the effects of emissions from chimneys, but additional coal consumption has a number of wider implications which will be considered at this conference.

Notably, it will entail the development of new deep mines and opencast sites (sometimes in areas with no mining tradition), with both social and environmental repercussions. It will result, too, in the movement of coal by road and by rail on a larger scale than hitherto and sometimes along non-traditional routes. It will also necessitate the provision of coal-handling and storage facilities (including port facilities) which have a visual impact on the environment.

Finally, there is the problem of dealing with the further large quantities of waste material that will result from mining and burning more coal. No doubt other speakers will be considering the question of colliery spoil, but perhaps I should mention the problem of disposing of extra tonnages of ash from power stations.

Not surprisingly the CEGB has pioneered the positive use of pulverised fuel ash. About 40 per cent of our output is used commercially. In particular, it has applications in concrete-making and it is used to manufacture lightweight building blocks. The handling and disposal of the remaining ash obviously require careful supervision and control if amenity problems for local communities are to be avoided. A great deal of this remaining ash is, in fact, put to beneficial use - for example as a fill for large areas of worked-out claypits at Peterborough, thus enabling derelict land to be reclaimed for farming, housing, or industry.

To sum up: Britain has a good record of controlling environmental pollution. In particular, the effects of burning coal in power stations are well known, and I believe we have adequate measures of control. But as new techniques of using coal arise, their environmental consequences will need to be examined with equal care.

The wider environmental and social implications of additional coal consumption will also have to be faced, and at the same time we shall need to guard against over-restrictive legislation.

This conference is a part of the process of careful consideration of all these issues, and I wish you every success.

Carcinogens from Coal and Other Sources

P. J. Lawther
MRC TOXICOLOGY UNIT, ST. BARTHOLOMEW'S MEDICAL COLLEGE, LONDON

The association of products from the incomplete combustion of fuels with the development of cancer can be traced back 200 years, to the original observations of Percivall Pott[1] on the occurrence of scrotal cancer among chimney sweeps. The wider implications of this finding were however recognised only slowly. In a review of the situation in 1892, Butlin[2] observed that the burning of soft coal in open grates appeared to be one of the factors responsible for the relatively frequent occurrence of chimney sweeps' cancer in Britain, and in 1922, nearly 150 years after Pott's observations, Passey[3] demonstrated experimentally the carcinogenicity of extracts of domestic soot. By then much evidence had accumulated, not only on the occurrence of scrotal cancer through contact with soot, but also on a wider range of skin cancers among workers exposed to coal tar products,[4] and there followed an intensive study of the carcinogenicity of tar fractions by Kennaway and his co-workers,[5] leading finally in 1933 to the identification and isolation of the potent carcinogen benzo(a)pyrene.[6]

Benzo(a)pyrene

Although it was realised then that the peculiarly British habit of burning coal inefficiently on open grates led to the dispersion of much tarry material into the air, the possible relevance of this to the development of lung cancer emerged only gradually as the incidence of this disease increased and as attention was drawn to the higher mortality in urban as compared with rural areas.[7] This fact was the stimulus for further studies on the potentially carcinogenic properties of coal smoke; benzo(a)pyrene was detected and determined in samples of domestic soot[8] and by 1952 the distribution of this compound had been examined in a number of towns in Britain.[9] The dominant source appeared to be the open coal fires that were still widely used, but contributions from industrial sources and motor vehicles were also recognised. Since that time there has been an intensification of interest in the role of carcinogens from these sources in relation to the growing problem of lung cancer throughout the world, but in view of the now well-established and clearly overwhelming effect of cigarette smoking on the development of this disease,[10] it may be time to pause and reflect on the findings to date.

Benzo(a)pyrene occurs together with a wide range of other polycyclic aromatic hydrocarbons in tar from the incomplete combustion of hydrocarbon fuels, and it is usually mixed with carbon and various inorganic components of the suspended particulate matter in urban air (referred to as 'smoke' in Britain). The range of polycyclic compounds that may be present, together with their structures and notes on nomenclature (benzo(a)pyrene was known earlier as 3:4-benzpyrene) and carcinogenicity has been well reviewed in a recent monograph.[11] There is no direct way of determining whether the mixture of pollutants, or any specific component of it, is carcinogenic to man when inhaled. The carcinogenic properties of the various polycyclic hydrocarbons present have been determined mainly from the results of experiments in which massive doses have been applied to the skins of animals. Benzo(a)pyrene is not the only component which, on this basis, is regarded as a potent carcinogen, but it is commonly taken as an index of that class of compound.

The contrast in concentrations that at one time existed between large cities, smaller towns, and rural areas, was the key to the interest then shown in the possible relationship between lung cancer mortality and benzo($\underline{a}$)pyrene in the air, and when measurements were made in Northern England as part of a large survey in 1958[12] a wide range of values was found (Table 1).

Table 1 Annual mean concentration of benzo($\underline{a}$)pyrene at sites in Northern England, 1958

	Benzo($\underline{a}$)pyrene $\mu g/1000 m^3$
Sites in major conurbations	
Salford (Regent Rd.)	108
Newcastle (Warncliffe St.)	75
Leeds	42
Liverpool (Edge Lane)	38
Other large towns	
Warrington	34
Burnley	32
York	24
Small towns	
Ripon	15
Wetherby	11

Large-scale surveys of that type have not been repeated in recent years, but if trends in smoke concentrations are taken as a guide it is clear now from the analysis of 10 years National Survey results[13] that throughout Great Britain there has been a large reduction in airborne polycyclic hydrocarbons during the 1960s, particularly in areas where concentrations were highest, resulting in a much reduced contrast between large and small towns.

Although there is no continuous series available to follow the declining concentrations of benzo($\underline{a}$)pyrene directly, results from several separate studies in London over a period of 25 years[9,14,15] have been collected in Table 2 to provide a further indication of trends.

Although there are differences in sampling site and in analytical methods within this series, the results suggest

Table 2 Concentrations of benzo(a)pyrene in air at sites in Central London, 1949-73, based on 24 hour samples aggregated for yearly periods

Period	Sampling site	Benzo(a)pyrene $\mu g/1000 m^3$
1949-51	County Hall	46
1953-56	St. Bartholomew's Hospital	17
1957-64	County Hall	14
1972-73	St. Bartholomew's Medical College	4

that the concentration of benzo(a)pyrene in the air is now only of the order of one tenth of what it was 25 years ago. If the earlier concentrations had been one of the factors related to the urban excess of lung cancer, then this substantial change might eventually be reflected in the mortality statistics. Since people dying now could still have been exposed to relatively high concentrations of smoke earlier in their lives, and since other factors apart from urban air pollution could be related to the urban/rural differences in lung cancer, the interpretation of current trends in mortality are difficult, but there are signs that the urban excess of lung cancer is declining.[16,17]

References

[1] P. Pott, Cancer scroti. In Chirurgical Observations, 1775, pp.63-68. Hawes, Clarke and Collins, London.

[2] H.T. Butlin, 'Cancer of the scrotum in chimney sweeps and others', Brit. Med. J., 1892, 1, 1341; 2, 1, 66.

[3] R.D. Passey, 'Experimental soot cancer', Brit. Med. J., 1922, 2, 1112.

[4] W.C. Hueper, 'Occupational tumours and allied diseases', pp.79-87, Charles C. Thomas, Springfield, Illinois, 1942.

[5] J.W. Cook, I. Heiger, E.L. Kennaway and W.V. Mayneord, 'The production of cancer by pure hydrocarbons', Proc. Roy. Soc. B., 1932, 111, 455.

[6] J.W. Cook, C.J. Hewett and I. Heiger, 'The isolation of a cancer producing hydrocarbon from coal tar', J. Chem. Soc., 1933, 395.

[7] P. Stocks, 'Regional and local differences in cancer death rates', H.M. Stationery Office, London, 1947.

[8] F. Goulden and M.M. Tipler, 'The identification of 3:4-benzpyrene in domestic soot by means of the fluorescence spectrum', Brit. J. Cancer, 1949, 3, 157.

[9] R.E. Waller, 'The benzpyrene content of town air', Brit. J. Cancer, 1952, 6, 8.

[10] Royal College of Physicians, 'Smoking and Health Now', Pitman, London, 1971.

[11] National Academy of Sciences, 'Particulate Polycyclic Organic Matter', N.A.S., Washington D.C., 1972.

[12] P. Stocks, B.T. Commins and K.V. Aubrey, 'A study of polycyclic hydrocarbons and trace elements in smoke in Merseyside and other northern localities', Int. J. Air Water Poll., 1961, 4, 141.

[13] Warren Spring Laboratory, 'National Survey of Air Pollution, 1961-71', H.M. Stationery Office, London, Vol. 1, 1972.

[14] B.T. Commins and R.E. Waller, 'Observations from a 10-year study of pollution at a site in the City of London', Atmos. Environ., 1967, 1, 49.

[15] B.T. Commins and L. Hampton, 'Changing pattern in concentrations of polycyclic aromatic hydrocarbons in the air of central London', Atmos. Environ., 1976, 10, 561.

[16] R.E. Waller, 'Tobacco and other substances as causes of respiratory cancer', in Symposium on the Prevention of Cancer, ed. R.W. Raven, pp.17-28, Heinemann, London, 1971.

[17] I.T.T. Higgins, 'Epidemiological evidence on the carcinogenic risk of air pollution', INSERM Symp. Series, 1976, 52, 41.

The Meteorological Effects of Increasing the Carbon Dioxide Content of the Atmosphere

Sir John Mason
DIRECTOR GENERAL, METEOROLOGICAL OFFICE, BRACKNELL, BERKSHIRE

Introduction
The concentration of carbon dioxide in the atmosphere has increased by about 15 per cent during this century and is currently rising at about 1/3% per annum as shown in Figure 1, largely owing to the burning of fossil fuels. In future the

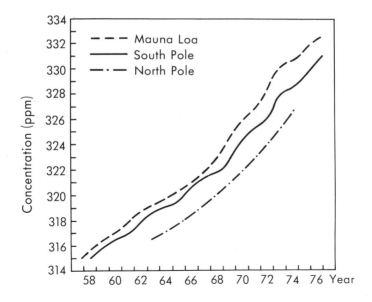

Figure 1   Measured increases in the concentration of atmospheric carbon dioxide since 1957

rate of rise will depend on future rates of consumption of these fuels and the rates at which the carbon dioxide released is assimilated by the earth's biota and by the oceans, none of which can be predicted with much confidence. If the present trend were to continue the CO_2 concentration would double from the present value of about 330 p.p.m. by volume to about 600 p.p.m. in about 200 years but calculations based on reasonable estimates of future fuel consumption suggest that such a doubling might well occur within 50 to 70 years - say between the years 2030 and 2050.

Since it strongly absorbs the infra-red radiation emitted by the earth's surface, higher concentrations of carbon dioxide should produce higher temperatures in the lower and middle atmosphere (troposphere) by the so-called 'greenhouse effect'. In addition, increased evaporation of water from the earth's surface as a result of the higher temperatures would increase the water vapour content of the atmosphere which, in turn, would absorb more of the terrestrial infra-red radiation and therefore reinforce the heating due to the increased carbon dioxide.

Early estimates of the warming due to increased CO_2 were exaggerated because they were based on simple, globally averaged, one-dimensional, single-column models of the atmosphere in which the enhanced downward flux of infra-red radiation was assumed to heat the earth's surface rather than the earth-atmosphere system as a whole, no allowance being made for the redistribution of the radiative heating by atmospheric motions. More sophisticated one-dimensional models, making some allowance for the vertical transport of heat by convective motions and for the radiative properties of water vapour and clouds, provide estimates for the globally averaged increase of surface temperature, due to a doubling of the CO_2 concentration to 600 p.p.m., ranging from 1.0 to 3.0 K depending upon the vertical distributions of temperature, humidity, and cloud cover. However, such simple models cannot properly represent the atmospheric dynamics nor possible feed-back mechanisms involving cloud, snow, and ice cover, and so the numerical results provide little more than a guide to the magnitude of the average heating effects with no indication of possible geographical variations. More realistic and detailed calculations require much more complex models involving a major effort by substantial teams of scientists equipped with computers of great memory capacity and speed.

Description of World Climate Models

Complex three-dimensional physico-mathematical models of the world climate have been developed in the Meteorological Office and by three major groups in the United States. These treat the atmosphere as a vast, turbulent, rotating fluid heated by the sun and exchanging heat, moisture, and momentum with the underlying continents and oceans and allow for monthly and seasonal changes in solar radiation, ocean-surface temperatures, ice and snow cover, vegetation cover, etc. They are based on the physical laws governing the changes in mass, momentum, energy, and water substance, on the equations of motion and continuity applied to a parcel of air, the laws of thermodynamics and radiative heat transfer, and the equation of state of a gas (air). A governing set of partial differential equations allows the atmospheric pressure, temperature, humidity, the three (E-W, N-S, and vertical) components of the wind, and their changes in space and time to be calculated.

The total mass of the atmosphere and its composition (fixed components), the rate of the Earth's rotation, the geography, topography (mountains), and albedo (reflectivity) of the Earth's surface are specified in advance. The incoming solar radiation at the top of the atmosphere is specified as functions of season and latitude and diurnal variations may be included. Sea-surface temperatures are usually held at their observed monthly/seasonal values but the temperatures of the land surfaces are determined by solving the relevant heat balance equations. Changes in soil and snow depth are computed as differences between precipitation (rain/snow), run-off, and evaporation. Changes in sea-ice cover may be computed but are held at their observed seasonal values in the Meteorological Office model.

Observed mean distributions of high-, medium-, and low-level clouds, now provided by satellites, are used to compute the transfer of solar and terrestrial radiation between different levels in the atmosphere. The radiation calculations also involve three absorbing gases - carbon dioxide, water vapour, and ozone. The mixing ratio of carbon dioxide is assumed to be constant everywhere. The distribution of water vapour is calculated step-by-step in the model but the variations of ozone with season, latitude, and height are specified by average observed values. The turbulent fluxes of heat, momentum, and water vapour at the Earth's surface and in the lower atmosphere are calculated in terms of their vertical gradients, the variation of wind and height, and the atmospheric stability. The transfer of heat and moisture by convective motions is also represented in simplified form. Condensation of water vapour is assumed to occur where the relative humidity of the air exceeds 100% and the excess moisture is deemed to fall out as rain or snow, allowance being made for evaporation if the precipitation subsequently falls through unsaturated air. The dynamical effects resulting from the release of the latent heat of condensation are automatically computed.

A serious weakness of present models is that the type and amount of cloud (as distinct from precipitation) are not computed so the important interactions between the clouds and the radiative field are not accurately represented although the effects of the <u>observed average</u> cloud cover on the radiation budgets are allowed for. A second major deficiency of the models is that they fail to represent interactions between the atmosphere and the oceans which, because they store and transport great quantities of heat, almost certainly exert a strong long-term control on the climate. Unfortunately we know relatively little about the deep ocean circulation and the changes in currents, temperature, salinity, etc. are not continuously monitored like atmospheric parameters.

Nevertheless, and despite these deficiencies, the best models successfully simulate the major features of the present world climate, at least as far as the averaged conditions are concerned. Having specified the boundary conditions, one may start from a given meteorological situation as represented by initial values of all the variables at a network of discrete points filling the whole of the atmosphere and then integrate the governing differential equations in short time steps over the period of many months or even years without the model becoming meteorologically or computationally unstable. Alternatively, one may start from an atmosphere at rest, from a motionless, isothermal, dry atmosphere containing no weather systems in which differential heating by the sun will produce pressure gradients and consequently air motions. As the integration proceeds, typically on a spherical grid with 5 to 20 levels in the vertical and some ten thousand grid points at each level, the wind gradually strengthens until transient eddies closely similar to the cyclones and anti-cyclones of the real atmosphere are generated. These redistribute the heat and moisture and the model, after some 100 days, reaches a state of statistical equilibrium and produces quantities which, when averaged over a month, reproduce quite realistically the distributions of temperature, pressure, wind, and rainfall observed in the real atmosphere. By changing the boundary of external conditions (e.g. solar input and sea-surface temperatures) month by month, the seasonal changes are also well simulated.

Maps showing the results of such simulations obtained with the Meteorological Office 5-level and 11-level global models may be found in Mason (1978, 1979).[1,2] These involve a great deal of numerical computation, a 100-day simulation calling for about 10^{12} arithmetical operations taking about 20 hours on our IBM 360/195 computer. With our new CDC 203(E) computer, to be installed early in 1981, we expect to increase these speeds 20-30 fold and be able to reproduce a whole annual climatic cycle in 2 to 3 hours.

Having demonstrated their ability to simulate the main features of the present climate, the models may be used with some confidence to investigate the response of climate to conceivable natural changes, for example in the sun's radiation, the land surface and vegetation cover, soil moisture, ocean-surface temperatures, etc. and to possible man-made changes to the carbon dioxide, ozone, dust, and heat content of the atmosphere and for judging whether these changes are likely to be distinguished from natural climatic fluctuations. Descriptions of several such model experiments will be found in Mason.[1,2] I shall now describe the results of the most recent attempt to model the climatic effects of increasing carbon dioxide.

The Effects of Increasing Atmospheric CO_2 in Three-dimensional Models

The only results from a three-dimensional model so far published are those by Manabe and Wetherald[3] using a very simplified limited-area model representing only about one-half of the northern hemisphere and excluding the polar regions above 82°N. The geography is highly idealized with very simplified continental shapes and no mountains. There is no diurnal or seasonal variation of the solar radiation, the clouds are prescribed as functions of latitude and altitude with no longitudinal variation, and the ocean is represented only by a wet-evaporating surface incapable of storing or transporting heat.

Starting from an isothermal atmosphere at rest, the model equations were integrated over a period of 800 days, the results being averaged over the last 100 days to give equilibrium climates for both the present concentration of carbon dioxide and for double this concentration. Doubling the carbon dioxide everywhere raises the temperature of the model troposphere and cools the stratosphere as shown in Figure 2. The increase in the average global surface temperature is 3 K, with a maximum increase of 10 K in polar regions caused partly by the retreat of the highly reflecting ice and snow surfaces and partly by the thermal stability of the lower atmosphere in these regions limiting convective heat transfer to the lower layers. In the tropics the heating is distributed through the entire depth of the troposphere by intense moist convection and so the temperature rise is smaller. Doubling the carbon dioxide also increases the intensity of the model's hydrological cycle, the average annual evaporation and precipitation both being increased by about 7%.

Mitchell[4] has used the much more complex Meteorological Office 5-level global model described earlier to investigate the climatic effects of doubling the present carbon dioxide content of the atmosphere. This model incorporates a fully interactive radiation scheme including diurnal and seasonal cycles, realistic geography and topography, cloud cover prescribed in terms of monthly averages derived from climatological records,

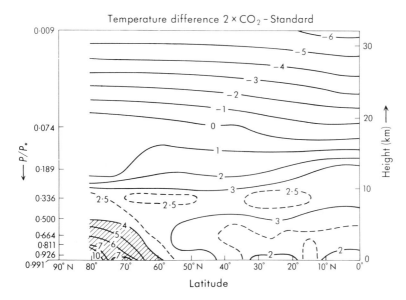

Figure 2 Zonally averaged increases in air temperature as functions of latitude and altitude due to a doubling of carbon dioxide to 600 p.p.m. as computed by Manabe and Wetherald[3] from a simple climate model.

and similarly derived sea-surface temperatures and sea-ice distributions updated every five days. Soil moisture is computed as the difference between precipitation, run-off, and evaporation; snow depth is computed as the difference between snowfall, melting, and evaporation and is then used in calculating the surface reflectivity. The model, starting from real observations on a particular day, was run for 500 days during which time it produced a realistic annual cycle.

Two sets of experiments were performed. In one of these, the model was run both with the present concentration of CO_2 and with double this concentration but the sea-surface temperatures were held at the same (normal) values in both. This is tantamount to assuming that the extra heat flux arriving at the ocean surface due to the increased CO_2 was immediately carried away by the ocean currents, none being stored in the ocean surface layers. Such a restriction would be expected to minimise the effects of increased CO_2 on the atmosphere. In fact doubling of the CO_2 concentration in these circumstances increased the mean global temperature of the troposphere by only 0.3 K and the annual mean land surface temperature by only 0.4 K. The mean global precipitation and evaporation actually fell by about 2.5% largely because the warming of the air over unresponsive oceans increased its stability and reduced the turbulent fluxes of moisture from the ocean surface to the atmosphere. Over land, where the surface temperatures were allowed to rise in conformity with the heat balance equations, increased radiative heating of the surface led to increased evaporation and precipitation, especially in summer when the average increase over the northern hemisphere continents was 2.4%.

In the second set of similar comparative experiments the ocean-surface temperatures were allowed to rise everywhere by 2 K in response to the additional radiative heat-flux resulting from the doubling of the carbon dioxide. This resulted in a rise of 2.2 K in the global annual average air temperature near the surface and a corresponding 5.3% increase in precipitation. Over the land masses alone the corresponding figures were 2.7 K and 3.2%. However, there were large seasonal, geographical, and regional variations as shown in Table 1 and in Figures 4 and 5.

The zonally averaged temperature changes as functions of latitude and altitude for the northern hemisphere winter and summer seasons are shown in Figure 3. These show that doubling the CO_2 content leads to a marked cooling of the lower stratosphere and a general warming of the troposphere with maximum heating in the upper troposphere and at low levels over the polar regions. This general pattern is similar to that obtained by Manabe and Wetherald[3] with their simpler model but, from the practical point of view, the magnitude, extent, and location of the anomalies may be more important than global or zonal averages. For example, a marked reduction in rainfall in a major food-producing region would be much more serious than a corresponding reduction in a non-productive region. It is therefore important to determine whether the spatial and seasonal variations in temperature and precipitation shown in

Table 1 The climatic effects of doubling atmospheric carbon dioxide with sea-surface temperatures allowed to rise 2 K

Northern Hemisphere Summer

	Northern Hemisphere Change due to x2 CO_2	Remarks
Land surface temperature	+ 3.0 K	Warming almost everywhere but max. rises up to 5 K at high latitudes and over S. Europe, Mediterranean, and Asia
Precipitation global land ocean	+ 5.8% + 3.4% + 7.0%	Slight reductions over Europe, Asia, and N. Africa
	Southern Hemisphere	
Land surface temperature	+ 2.7 K	Mostly warming over continents, including Antarctica
Precipitation global land ocean	+ 6.5% − 2.6% +11.3%	Reductions over parts of Africa, S. America, and Australia

Northern Hemisphere Winter

	Northern Hemisphere	
Land surface temperature	+ 2.6 K	Max. warming of up to 10 K over N. Europe and Asia
Precipitation global land ocean	+ 4.6% − 5.8% + 9.2%	Reductions mainly over USA, Europe, and EuroAsia mainly in tracks of middle-latitude depressions
	Southern Hemisphere	
Land surface temperature	+ 2.9 K	Max. warming over Antarctica up to 4 K
Precipitation global land ocean	+ 9.0% 12.5% 6.7%	Max. increases over eastern sides of sub-tropical continents

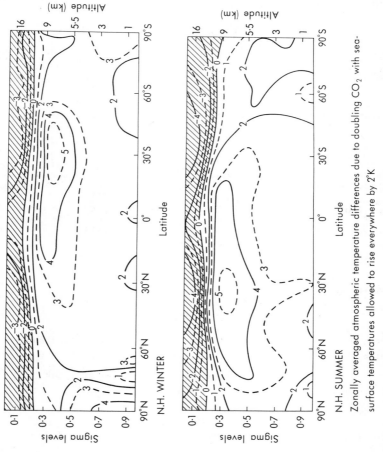

Figure 3 As for Figure 2 but calculated from Meteorological Office global climate models: (a) Northern Hemisphere Winter; (b) Northern Hemisphere Summer. Zonally averaged atmospheric temperature differences due to doubling CO_2 with sea-surface temperatures allowed to rise everywhere by 2°K

Figures 4 and 5 are 'real' in the sense that they are statistically significant relative to the natural variability or 'noise' of the model. To this end the changes ascribed to a doubling of the CO_2 have been subjected to statistical tests and the changes $\geq$ 2 mm day^{-1} in Figures 4 and 5 are estimated to be significant at the 10% level, i.e. there is less than 10% probability that the anomaly would occur as a result of random model fluctuations. In this context the temperature changes shown in the diagrams appear to be statistically significant but only the larger changes of rainfall confined to limited areas such as those on the eastern sides of the sub-tropical continents in summer can be clearly distinguished from the model variability. This is not to say that such changes would necessarily occur in the real atmosphere if the carbon dioxide were doubled, for reasons that we shall discuss later, but they are indicative of the kind of changes to be expected.

Recently Manabe and Wetherald[5] have carried out some new computations with a rather more realistic version of their earlier simple model which retains the very idealised geography but now extends to the Pole but still has no mountains, no seasonal or diurnal variations of radiation, and no ocean storage or transport of heat. However, the cloud is now computed by an elementary scheme instead of being prescribed. Starting from an isothermal, dry atmosphere, long-term integrations spanning 1200 days were carried out for CO_2 concentrations corresponding to one, two, and four times the present value and the resulting climates were obtained by averaging over the last 500 days of each integration.

Figure 6 shows the latitude-height distributions of the changes in zonal-mean temperature for two- and four-fold increase in CO_2. The results for doubling of CO_2 are very similar to the earlier results of Manabe and Wetherald,[3] the average rise in surface temperature over the whole domain being 3 K with a maximum of 8 K near the Pole. The corresponding figures for quadrupled CO_2 are 5.9 and 15 K. The rise in temperature is accompanied by a marked increase in the water vapour content of the lower atmosphere leading to poleward transports of latent heat at middle latitudes and increased

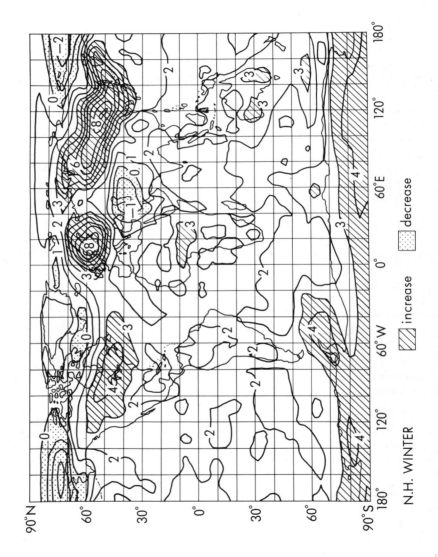

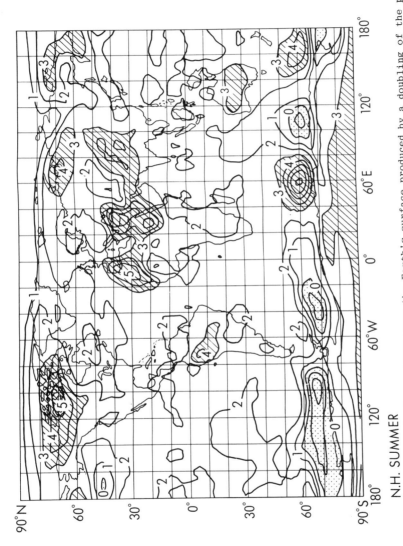

Figure 4 Changes in air temperature near the Earth's surface produced by a doubling of the present concentration of carbon dioxide in the Meteorological Office climate model when the sea-surface temperature is allowed to rise everywhere by 2 K: (a) Northern Hemisphere Winter; (b) Northern Hemisphere Summer.

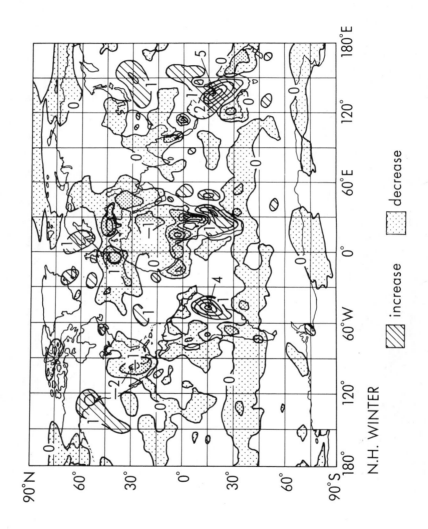

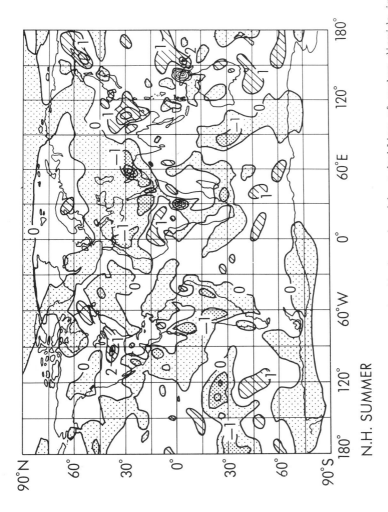

Figure 5 Changes in precipitation (mm/day) produced by a doubling of carbon dioxide in the Meteorological Office model: (a) Northern Hemisphere Winter; (b) Northern Hemisphere Summer.

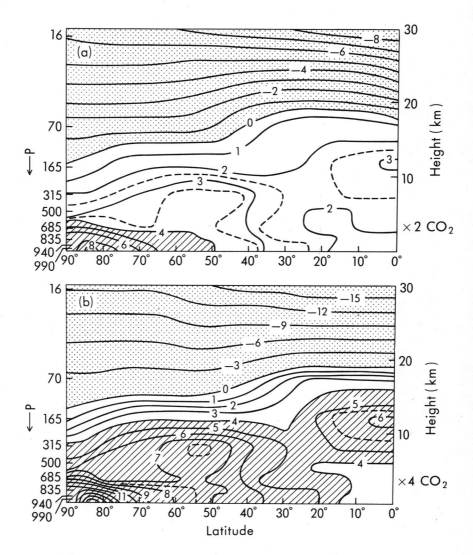

Figure 6   Changes in atmospheric temperature as functions of latitude and altitude produced by a doubling and quadrupling of carbon dioxide in the modified simple model of Manabe and Wetherald:[5]   (a) doubling of CO_2; (b) quadrupling of CO_2.

precipitation. Doubling the CO_2 led to an average increase in precipitation over the whole area of 7%, again in close agreement with the earlier work; for a four-fold increase in CO_2 the precipitation increased by 12%.

Even in such a simple model the changes in precipitation were far from uniform and, like the Meteorological Office model, showed marked geographical variations. Increasing the CO_2 caused a marked poleward shift of the middle-latitude rainbelt, a reduction in soil moisture over the continents at around 40° latitude, and an intensification of the monsoon rains along the eastern coasts of the model continent.

Manabe and Stouffer[6] have also used a nine-level, three-dimensional global model, similar to the Meteorological Office model, to investigate the effects of quadrupling the carbon dioxide but representing the oceans by a well-mixed static layer (no lateral heat transport) everywhere 68 metres deep and with provision for a layer of sea ice. In the presence of sea ice the underlying water is held at freezing point and the heat flux through the ice is balanced by the latent heat of freezing/melting at the bottom. This, together with the snowfall, melting, and evaporation on the top surface determines the thickness of the ice.

After running for about 10 years the model settles down into a stable climate in which the global mean temperature of the ocean layer increases from 293 to 296 K* when the CO_2 is quadrupled to 1200 p.p.m. The average annual increase in air temperature near the surface of the Earth increases by 4.1 K, the increase being greater (4.5 K) in the northern hemisphere than in the southern hemisphere (3.6 K) because of the greater land mass. In low latitudes the additional warming is small and varies little with season, but in high latitudes it is generally larger and varies markedly with season especially in the northern hemisphere from +14 K over the polar cap in early winter to only +1 K in mid-summer. In middle latitudes (50°N)

---

* This result is fairly consistent with Mitchell's assumption that a doubling of CO_2 would lead to a rise of 2 K in ocean surface temperatures.

it is fairly constant at 5-6 K at all seasons. The warming is much less at high latitudes in summer because the sea ice is absent or thinner and the increased radiative flux due to increased CO_2 is used largely in melting the upper surface of the ice or warming the ice-free mixed ocean layer rather than in warming the overlying air.

Comparison of Model Results

The results of the various three-dimensional models, which differ a good deal in complexity and in their detailed formulation, are nevertheless in rather good agreement at least as far as the effects of doubling the present levels of CO_2 on the global annual averages of surface air temperature and precipitation are concerned. The estimated increases lie in the ranges 2-3 K and 5-7% respectively. In all models maximum heating occurs in polar regions in winter, the model estimates ranging from 7 to 10 K. The largest increases in precipitation tend to occur in middle latitudes between 40 and 70°.

However, as already pointed out, the geographical and regional variations may be of greater economic and political significance than globally averaged values. It is therefore important to assess the 'reality' of these variations and major anomalies, to judge whether they are likely to occur in the real atmosphere or whether they are peculiarities of the particular model. For however impressive the models may be in reproducing the broad features of the natural global climate, it is important to remember that they are still gross simplifications of the real atmosphere/ocean climatic system and can hardly be expected to simulate its fine detail.

In order to make these assessments and to narrow the degree of uncertainty in the model predictions we have to improve and refine the physics and dynamics of the models, compare the results of different models, and test the sensitivity of the results to changes in the model parameters and see to what extent the predicted changes still remain detectable above the model noise. This will call for a long and hard effort which can be undertaken in only a very few centres because of the human and computing resources required. However, we may take some comfort from the fact that modest changes in temperature and precipitation brought about by increasing CO_2 are likely to have some beneficial effects as well as some possible disadvantages. A general warming will help reduce fuel consumption while higher temperatures and rainfall will benefit agriculture (apart from the direct benefits of the increased CO_2) at least in some parts of the world. Finally we shall describe a very interesting and surprising additional benefit that may arise from an increase in the carbon dioxide content of the atmosphere.

Effects of Increased CO_2 on Stratospheric Ozone

Carbon dioxide is not directly involved in ozone chemistry but influences it through its effect on stratospheric temperatures. We have seen earlier that, because the carbon dioxide in the upper stratosphere emits more infra-red radiation to space than it absorbs, an increase in its concentration cools the upper stratosphere where most of the ozone is formed.

Groves *et al.*[7] computed these effects by following the progress of 28 simultaneous reactions thought to be of the greatest importance in ozone chemistry all assumed to take place in a vertical column at 34°N containing a realistic vertical distribution of water vapour and a single layer of cloud reflecting 30% of the incident solar radiation. Diurnal and seasonal variations of solar radiation are included and rates of vertical turbulent mixing were taken from the Meteorological Office 13-level global three-dimensional model but otherwise dynamical effects are ignored. The vertical temperature profile is determined by calculating heat transfer between levels by both radiative exchange involving all the absorbing gases and by convection. The temperature profile is updated in the chemical kinetics scheme every 30 min and the resulting ozone profiles are used to update the radiative heat transfer calculations.

Figure 7 shows the computed changes of temperature and ozone concentration caused by increasing the CO_2 concentration from 290 to 600 p.p.m. At heights of 40 km the temperatures are lower by about 10 K in summer and about 7 K in winter, the corresponding increases in ozone being 14 and 11%. At 30 km

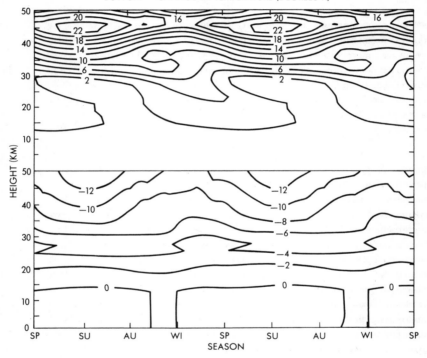

Figure 7 Computed changes in temperature and ozone concentration as functions of altitude and season due to a doubling of CO_2 concentration to 600 p.p.m. (Groves et al.)[7]

the temperature changes are respectively −6 K and −4 K, in good agreement with Figure 2, but the ozone increases by only 4% in the summer because the large increase at higher levels absorbs much of the incoming ultra-violet radiation and allows less to penetrate to the 30 km level.

The total integrated ozone content of the vertical column is estimated to increase by 5.5% as the result of doubling the CO_2 to 600 p.p.m. and this may largely offset any reduction in ozone caused by the continued release of chlorofluorocarbons (CFMs) from aerosol sprays and refrigerators. Groves and Tuck,[8] using the model just described, calculate that if CFMs continue to be released at the 1975 rate for the next 50 years, they, together with natural chlorines in the atmosphere, will reduce the total ozone by 8.6%. This could well be largely offset by a doubling of CO_2 over the same period so that the overall reduction in ozone would be only 2.6% (the effects are not quite additive) which would be difficult to distinguish from the large natural spatial and seasonal variations in ozone.

Conclusions

Although we can make only an educated guess at the rate at which atmospheric carbon dioxide is likely to increase in the future, largely because we neither know nor understand the rates at which the CO_2 released by the continued burning of fossil fuels is likely to be taken up by the oceans and biosphere, it is quite likely that its concentration in the atmosphere will double to 600 p.p.m. during the next 50 years or so. In this case, the best available climatic models are in fair agreement in predicting that the global annual average surface air temperature will rise by 2-3 K with maximum warming of up to 10 K in polar regions in winter. The corresponding increase in precipitation will be 5-7% with considerable geographical and regional variations, the largest increases of up to 20% tending to occur in middle latitudes. However, these are only tentative estimates obtained with simplified and imperfect models in which the important roles of the oceans in storing and transporting heat and the interactive effects of clouds and the polar ice caps are not satisfactorily represented.

An increase in atmospheric carbon dioxide is also likely to lead to a cooling of the stratosphere and an increase in ozone at these levels which may largely offset the destruction of ozone by the continued release of chlorofluorocarbons from aerosol cans and refrigerators.

All these changes, which will occur only slowly over several decades, are likely to have some beneficial and some detrimental effects, the overall balance between which cannot be properly assessed until the predictions of the climatic models attain rather greater precision and reliability. Meanwhile there is no good reason to believe that increasing carbon dioxide will have dramatic or disastrous consequences for the global climate during the next 50 years; the effects are likely to be marginal overall but possibly rather more serious in certain localised areas. Much more research will be needed before we can say with confidence what and where these are likely to be.

REFERENCES

[1] B.J. Mason, Proc. Roy. Soc., 1978, A363, 297-333.
[2] B.J. Mason, Proc. World Climate Conference (World Meteorological Organization, Geneva), 1979, 210-242.
[3] S. Manabe and R.T. Wetherald, J. Atmos. Sci., 1975, 32, 3-15.
[4] J.S. Mitchell, 1979, in preparation.
[5] S. Manabe and R.T. Wetherald, J. Atmos. Sci., 1980, in the press.
[6] S. Manabe and R.J. Stouffer, Nature, 1979, 282, 491-492.
[7] K.S. Groves, S.R. Mattingley and A.F. Tuck, Nature, 1978, 273, 711-715.
[8] K.S. Groves and A.F. Tuck, Nature, 1979, 280, 127-129.

Discussion on Session I

J G Collingwood (Commission on Energy and the Environment): Is it still the opinion of the CEGB that on balance it is better for the environment to refrain from washing SO_2 out of flue gases and rely only on high-temperature discharge at high level?
G England: Yes. Our experience at Battersea and Bankside Power Stations indicated that gas-washing results in the production of considerable amounts of liquid that are not easy to dispose of. In addition gas-washing cools the plume so that it is less readily dispersed.

B Lees (Consultant, Institute of Energy): I agree with Mr England that the combustion of coal in power station boilers instead of in the open fire in homes and offices played a part in reducing the smoke problem since the Clean Air Act 1956. Has he not failed to mention, however, the important influence of the discovery of and change to natural gas and the application of oil-fired equipment for heating purposes in homes, commercial premises and industry as a major factor in reducing air pollution, particularly as these forms of heating are more efficient than electricity?
G England: In my view the main contribution to the smoke problem was made by the creation of smokeless zones and the disappearance of the open fire, but I agree that the other factors were contributory.

J Rowcliffe (Central Directorate on Environmental Pollution, Department of the Environment): Mr England rightly drew attention to the fact that policy must take account not only of SO_2 concentrations in this country but also of international (transboundary) effects. There is increasing international activity on this issue - a proposed EEC Directive and pressure for an Emissions Policy: the recent EEC meeting at which a Convention on Transboundary Air Pollution was signed: and wider interest in view of the recent evidence that European emissions may affect not only Scandinavia but the Arctic. Against this background the scientific basis for the alleged effects of 'our' SO_2 on other countries is important; and could Mr England enlarge on the doubts that have recently been cast on the evidence for such effects?

G England: I agree the importance of establishing the scientific basis of any claim for adverse effects of pollution and not embarking prematurely on abatement policies that may or may not be justified. In this connection the more cautious approach in the UK has avoided the substantial costs unnecessarily imposed on industry in USA and in parts of Europe in advance of scientific investigations, which in the event failed to justify the need. Subsequently speakers at this Conference will describe the specific areas of uncertainty in the evidence for adverse effects from long-distance transport of sulphur dioxide.

D W Gill (National Coal Board): (Question to Sir John Mason): There have been very considerable changes in land cover during this century, resulting from man's activities, for example the spread of deserts due largely to overgrazing, and the destruction of tropical forests. Has the Meteorological Office model given any indication of the magnitude of the effects of this on the world's climate?
Sir John Mason: We have not investigated the destruction of tropical forests or large-scale overgrazing but we have used the climate model to investigate the effects of making a large part of the Sahara wet instead of dry. The general effect is to produce more vigorous depressions fed by the evaporation of

moisture from the underlying wet soil and this in turn produced much higher rainfall and so tended to preserve the soil moisture. In other words wetting the surface on this large scale would probably tend to persist and I would almost expect a major extension of the desert to be self-generating.

Session II

Chairman's Opening Remarks

R. Press
CHAIRMAN, COUNCIL OF SCIENCE AND TECHNOLOGY INSTITUTES

Introduction
As Chairman, CSTI, I fully subscribe to the statement of purpose of this conference as set out on back of the programme card and I believe that CESE should be congratulated on its initiative and timeliness in arranging it.

I think, however, that it is worth making more explicit two points which are certainly implicit in the statement of purpose:

(i) It is not only that an authoritative review of the consequences of increased burning of coal would be of 'considerable interest to engineers, scientists, etc. -' but also that there is a real public need for authoritative and objective information on whatever hazards there may be so that public opinion and judgements may be based, as far as possible, on equally well informed assessments of the environmental effects of alternative sources of essential energy supplies.

(ii) The second is closely related to the first in that comparative lack of interest, so far, may well have been merely because the exercise now initiated by this conference has not already been pursued to a depth comparable with that, e.g., to which nuclear energy has been exposed.

I believe the papers being presented here today go right to the heart of what is now needed. One is here reminded of the title of an article in a recent issue of The New Scientist (25.10.79): 'Coal takes its turn in the dock'!

The General Problem

The risk to human health is obviously one of many factors which must affect our decisions on the suitability of different sources of electricity production. A difficulty is that no single index of harm - whether based on total number of lives lost, or length of life lost, or any other simple numerical criterion - will necessarily give a measure of general public acceptability, still less of the views of different individuals.

Another difficulty is when there is inadequate information from human epidemiological surveys of the frequency with which a pollutant will cause harmful effects in man. In this respect, the effects of discharges from the combustion of fossil fuels are much less well quantified than are those from nuclear reactors. The effects of the combustion products of fossil fuels are much less well known, partly because of the difficulty in making well controlled epidemiological surveys, partly for lack of any adequate counterpart for the long and thorough study of radiobiology, and partly also because no single component of the discharges has been shown to give an index of harmfulness in the way that the absorbed dose of ionising radiation is likely to do. In the absence of information leading to estimates of objective risks, and assessment of course of action likely to minimise the amount of human harm, public opinion on the choice of alternatives will be more likely to be based on attitudes towards the types of risk than on the size of the risks.

Mr. A.J. Clarke
Head of Environmental Section, Planning Department,
CEGB Headquarters

An Honours graduate in mechanical engineering of Loughborough Technical College, Mr Clarke was commissioned in the Transportation Branch of the Royal Engineers and served in Port Operations in Egypt.

He joined the CEGB (then BEA) in 1949, serving in a number of operating and maintenance posts in power stations before transferring to Board Headquarters in 1952, to join Station Planning Section. Increasing specialisation in the environmental and pollution control aspects of power station planning led to his appointment in 1958 to a new group of specialists within the Planning Department, formed to deal solely with these matters. He was appointed Head of the Environmental Group in 1961, where he has been responsible for initiating a number of technical advances, including the use of single multi-flue chimneys for major plants; for planning large-scale pollution monitoring surveys; and for advising the Board generally on environmental policy. Following an expansion of the Group's activities, Mr Clarke was appointed Head of the Environmental Section in 1978.

Mr Clarke has written many technical papers, principally on air pollution control and has represented both the Generating Board and the United Kingdom on a number of international conferences and delegations. He has also undertaken many overseas consultancy services on behalf of the Generating Board.

Dr. M.W. Holdgate, FIBiol
Director General of Research, Department of the Environment and Department of Transport

Dr Holdgate is a biologist by training and a specialist on the effects of pollution. He has been a University lecturer, the Chief Biologist of the British Antarctic Survey, the Deputy Director of the Nature Conservancy, the Director of the Institute of Terrestrial Ecology (Natural Environment Research Council), and the Director of the Central Unit on Environmental Pollution in DOE. He has held his present post since 1976.

Chairman's Opening Remarks

Dr. D.H. Napier, MSc
Senior Lecturer in Industrial Hazards, Department of Chemical Engineering, Imperial College of Science & Technology

Dr Napier is also Chairman of the College Safety Council, Chairman of the Explosion and Related Hazards Committee, Fire Research Station, Borehamwood, and Chairman of the University of London Safety Officers' Committee. He was visiting Professor at the University of Toronto from September to December 1979.

From 1951 to 1957 he was Scientist in charge of the Combustion Department of the British Coal Utilisation Research Association and from 1957 to 1965 he was the Head of the Chemical Projects Section at Vickers Research Limited. He joined Imperial College in 1965.

Emission of Sulphur Dioxide, Nitrogen Oxides, and Particulate Matter from Coal-burning Power Stations

A. J. Clarke
PLANNING DEPARTMENT, CENTRAL ELECTRICITY GENERATING BOARD, LONDON

In the course of this paper I shall be able to deal only briefly with some of the more important aspects of my subject that fall within the main theme of this conference. I shall deal firstly with SO_2 emissions as this illustrates many of the problems and contentious areas in respect of control policy, and then rather more briefly with NO_x and solid emissions.

To set the scene, however, I will start with some technical facts. Figure 1 shows the emissions rates for sulphur dioxide (SO_2), nitrogen oxides (NO_x), and particulate matter from a typical 2000 MW coal-fired power station, based on average UK coal quality and operating conditions. SO_2 has the largest emission rate, 6.9 kg s^{-1} at maximum station output or 120,000 tonnes per annum at 56% load factor. NO_x emission is less than one quarter of this by weight and the particulate emission is only 3% of the SO_2 emission. Some of these figures may seem frighteningly large - particularly the annual emissions - and, in the remainder of this paper, I shall describe their impact on the environment. But I would like to begin with a general discussion of environmental standards and controls in relation to air quality.

Figure 2 illustrates a sequence of events, beginning with the fuel itself and ending with some particular impact in the environment. In between, the fuel is burned in the plant and the flue-gases discharged to atmosphere. They then disperse in the atmosphere until they reach a receptor zone where they cause - or may potentially cause - some adverse impact. At the different stages along the way control standards can be set and control techniques applied. At the first stage, for instance, a fuel quality standard may be set, such as a

TYPICAL EMISSION RATES FOR 3×660 MW COAL FIRED STATION :

	CONCENTRATION IN FLUE GAS	EMISSION RATES	
		Kg/sec @ MCR	Kt/year @ 56% L.F.
SO_2	1200 vpm	6.9	120
NO_x	550 vpm	1.5	26
PARTICULATES	115 mg/m^3	0.22	3.9
COAL CONSUMPTION		215	3,800

ASSUMPTIONS :
CALORIFIC VALUE 24,930 KJ/Kg
SULPHUR CONTENT 1.6 %
ASH CONTENT 16.4 %
PRECIPITATOR EFFICIENCY 99.3 %

Figure 1

limitation on the sulphur content of the fuel that may be burned. The related techniques include selection of fuel naturally low in sulphur or pre-treatment to remove part of the sulphur. Alternatively, post-treatment might be used, such as gasification coupled with desulphurisation of the resulting gas.

Standards and controls can be applied to the plant itself or at the point of emission. The latter, however, are more concerned with ensuring dispersion in the atmosphere after emission - an adequate stack height, for instance, and means of minimising aerodynamic disruption of the plume. A point to which CEGB attaches great importance is the use of a single

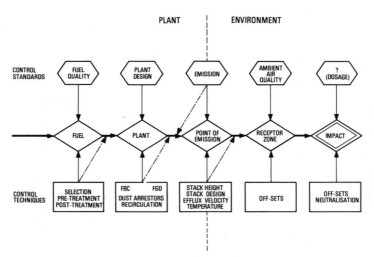

ENVIRONMENTAL STANDARDS AND CONTROLS - AIR QUALITY

Figure 2

chimney to maximise the thermal rise of the plume. After dispersing in the atmosphere, some of the flue-gas will return to ground level in a receptor zone, where the appropriate standard is ambient air quality, e.g. the maximum permissible concentration of, say, SO_2 averaged over a day or a year. Some measure of control by off-sets is still feasible; e.g. the contribution to ground-level concentrations from a new source must be counter-balanced by reductions from an existing one. A similar policy is adopted in the USA but there it is applied to mass emissions instead of ambient concentrations, which is altogether a different approach and less effective.

The presence of some diluted flue-gas in the receptor zone may lead to an adverse impact, on human health, on growing crops, on corrosion rates, etc. There are very few examples of control standards applied at this point despite the fact that this is the real focus of attention in the whole chain of events. The purpose of this explanation is to draw some relevant conclusions, the first of which is that control standards become progressively more difficult to apply - in terms of the ease in setting them and ensuring that they are maintained - as one moves closer to the vital target, which is the point of impact in the environment. Almost a corollary of this is that our scientific knowledge is less as we move in the same direction. A great deal is known about the presence of pollutants within the fuel and how they behave in the plant and we have a reasonable working knowledge of the diffusion process in the atmosphere, but we are badly in need of more objective knowledge about the impacts caused by these pollutants when they reach the environment, usually in very diluted quantities.

Both of these limitations encourage the setting of standards and controls further back in the chain of events, but this introduces an opposing factor of uncertainty. The more remotely the controls are applied from the point of impact, the less positive is their action and the less certain their effectiveness. Controls applied at the front end of the process tend to be arbitrary and draconian. They may limit the impact, as desired, but possibly at the expense of drastically over-controlling the whole process in a way that is far from cost-effective. This situation is a dangerous one, and it underlies most of the international controversy that arises over methods of controlling pollution. It can only be resolved by acquiring a more exact knowledge of the actual impacts that SO_2 and other pollutants cause in the environment so that, if necessary, appropriate standards can be devised for the point where they are most effective.

Now I shall look rather more closely at Figure 3 and consider effects at different distances from the source. Keeping to the example of SO_2 emission, this may reach receptor zones over a wide range of distances. Short-range effects, as shown in the Figure, are predominantly those on the concentration of SO_2 at ground level. The approximate order of distance is about 10 km and the appropriate control standard is ambient air quality, as before. At middle

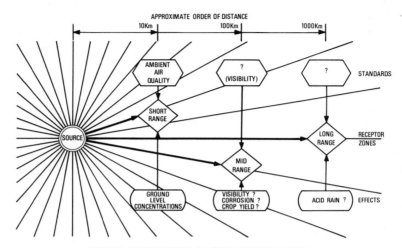

EFFECTS OF EMISSIONS AT DIFFERENT DISTANCES

Figure 3

distances, of the order of 100 km, some of the SO_2 may be converted into sulphates, which may affect atmospheric visibility or enhance the corrosion rates of metal and stone. The unconverted SO_2 is by now widely dispersed, but may still have an impact on the growth of plants. There are no practical standards yet proposed for these impacts but I believe standards based on visibility are under discussion in the USA. At long distances, of the order of 1000 km, the principal impact is on the acidity of rainfall, about which I shall have a good deal more to say later. Again, the uncertainties in our present knowledge preclude the adoption of standards for long-range effects.

The background pattern of radial lines in Figure 3 is not merely artistic licence; it illustrates my previous point about the difficulties of controlling pollutants remotely from the point of impact. The further away the receptor zone from the source, the smaller the angle it subtends and the more unlikely it is that emissions from that source will cause any impact.

I shall deal first with short-range emissions. A number of countries have adopted ambient air quality standards, or guidelines. The SO_2 standards chosen, illustrated in Figure 4,

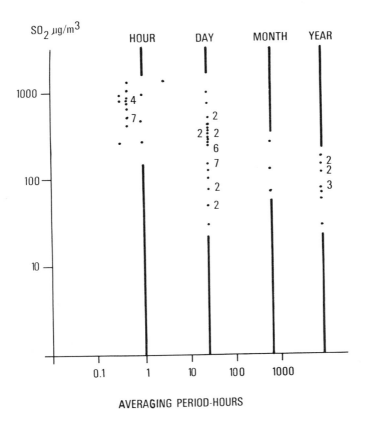

AIR QUALITY STANDARDS FOR SO_2

Figure 4

are depicted by dots against the appropriate sampling time. The numbers indicate cases where the same standard has been chosen by several different countries. Note that the scales are logarithmic so that the disparity in the standards chosen is even greater than it appears at first sight. The daily-mean standard, for instance, varies from 30 to 1000 µg m^{-3}, a factor of 33 to 1. Reading across the Figure, too, it can be seen that one country will accept, as an annual average, an SO_2 concentration nearly as high as another country will tolerate for 15 minutes only. This is rather a dramatic illustration of the dearth of hard factual data on the adverse effects of SO_2 as a pollutant. Presumably, the criteria on which all these standards were based is the same body of internationally available information, for instance, on the effects of SO_2 on public health, on vegetation, etc. These data are clearly open to widely different interpretations even at the scientific level, which in turn allows political expediency to dictate where the standards should be set.

Turning now for comparison to the actual impact of a modern coal-fired power station, the CEGB has built up a considerable body of data over the last 30 years as a result of many field surveys and research studies. Figure 5 shows the results of one such study at a 2000 MW coal-fired station. It gives the frequency with which particular SO_2 concentrations were exceeded for various averaging periods: three minutes, one hour, one day, etc. Background concentrations have been deducted and the frequencies normalised for wind direction. The curves therefore represent the impact of the power station alone at the radius of maximum effect and at a point that receives an average proportion of winds from the station. For instance, an hourly mean concentration of 20 µg m^{-3} - the average background level in clean rural areas in the UK - was exceeded for less than 5% of the time, and for the remaining 95% of the time the station contribution was less than the background, or was actually zero. Over the course of a year, the effect of the station was to increase the background by only 2 or 3 µg m^{-3}. Compared with the standards shown in Figure 4, the power station can comfortably comply in all cases except the few that are inordinately severe.

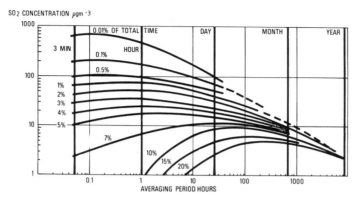

FREQUENCY DISTRIBUTION OF SO₂ CONCENTRATIONS FROM 2000MW POWER STATION AT RADIUS OF MAXIMUM EFFECT (~8Km) - NORMALISED FOR WIND FREQUENCY

Figure 5

These curves were derived from several million individual readings taken with an array of 16 continuous recorders and thus very accurately represent the true impact of the station which, at the time, was burning coal of about 2% average sulphur content, and emitting rather more SO_2 than the typical figures given in Figure 1. The power station at which this study was made was Eggborough, one of a group of large stations in Yorkshire. Together with Ferrybridge and Drax, this makes a total of over 6000 MW of coal-fired generating capacity within a distance of 20 km along the Aire and Ouse valleys. The contours shown in Figure 6 are the annual SO_2 concentrations for a particular year, derived from the National Survey of Air Pollution gauges in the area. The gradient of pollution is dominated by the dense industrial and residential areas of west Yorkshire and, to a noticeable extent, even by the smaller communities in the more agricultural areas. It is not evident, however, that the

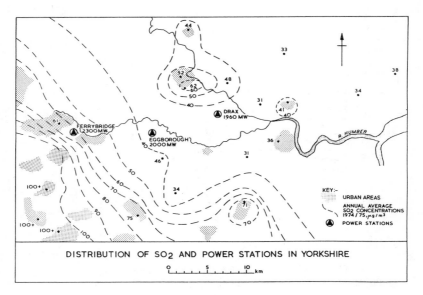

Figure 6

contours are influenced at all by the presence of the power stations which collectively emit far more SO_2 than the rest of the area combined. Both this study and others show that a power station's SO_2 emission is virtually undetectable on the standard daily gauges of the National Survey.

As regards the mid-range dispersion of SO_2, there is very little firm information to be given. This is an area which, at first, aroused little interest and subsequently has been rather overshadowed by the attention given to long-range effects and acid precipitation. Some studies on the growth of crops have suggested that relatively modest concentrations of SO_2 can depress the yield without visible injury to the plants. However, the application of these studies to practical field conditions is difficult; many were undertaken in the artificial conditions of growth chambers and at constantly maintained SO_2 concentrations which turn out to be more persistent than those actually experienced, even in city centres. More research needs to be done before it can be judged whether or not power station emissions can measurably affect crop yields, bearing in mind that these emissions are unlikely to add 10%, at the most, to rural background SO_2 concentrations in this country.

SO_2 is converted into sulphates in the atmosphere fairly slowly, often combining with ammonia to form a very fine ammonium sulphate aerosol. The resulting haze may reduce the clarity of the air, slightly but over wide areas. The chemical processes are not fully explained but it is reasonable to expect that they involve the total SO_2 burden in the atmosphere, not just that from power stations. The disbenefit of a slight reduction in visibility is rather a subjective issue, but this is not to say that it is unimportant. Further research is needed.

Corrosion of metals and stonework is also an area of doubt where the contribution from power stations cannot yet be defined. An unpolluted atmosphere is, of course, still corrosive as a result of moisture containing dissolved carbon dioxide. The great limestone caverns of the world were created by carbonic acid rain long before mankind made his appearance. The addition of sulphuric and nitric acids from fuel combustion no doubt accelerates the corrosion process but the extent to which it may be alleviated by control at source, and the ratio of costs and benefits, is still a matter of conjecture.

The long-range transport of pollutants over distances of the order of 1000 km is, of course, of great topical interest, and a subject to which the more popular technical journals and the press are devoting zealous attention. Most authors assume that the causal connection between distant SO_2 emissions and, say, the disappearance of fish species from remote lakes has been firmly established and that, furthermore, the situation is steadily getting worse, year by year. However, the stridency of many of these publications tends to conceal a number of fundamental uncertainties about acid rain that fully justify a more cautious approach.

First, let us look briefly at the dispersion factors involved (Figure 7). SO_2 from a particular source diffuses until it occupies the atmospheric mixing layer, which extends 1 or 2 km above the surface. The SO_2 may be lost by dry deposition -

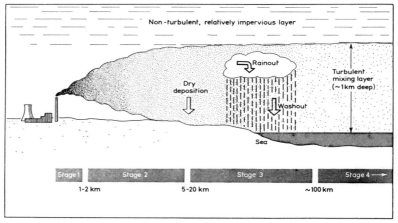

The four main stages of chimney-plume dispersion

Figure 7

that is, direct absorption as a gas at land and sea surfaces - or by wet deposition - where the gas may be dissolved in cloud droplets during the rain-forming process (rain-out) or be scavenged from the lower atmosphere by falling rain (wash-out). Solution of sulphuric and nitric acids in rain will, of course, increase its acidity.

One popular misconception can be dispelled immediately. After a relatively short travel distance - 20 to 100 km - the SO_2 is fully mixed and it is no longer possible to distinguish contributions from tall stacks and from ground-level sources. Indeed, the mathematical model of long-range transport developed during the recent OECD study made the simplifying assumption that the mixing took place instantaneously after emission, without material loss of accuracy. Any problem with acid rain is, therefore, not specifically related to tall stacks but is related to total emissions from the source area.

Mathematical models, like that just mentioned, can be used to calculate the rate of sulphur deposition from a particular source area, using relevant emission and meteorological data. Figure 8 shows the total sulphur deposition (wet plus dry) over Western Europe from UK sources alone for a particular year. A more elaborate model can calculate the combined deposition from all countries, shown in Figure 9. These two Figures dispel one or two other popular misconceptions. Critics of tall stacks often allege that they are, essentially, a selfish technique in that they preserve the local environment at the expense of that further away. 'Dumping our rubbish in someone else's backyard' is a phrase often heard. This is not the case; sulphur deposition is greatest in the source area and rapidly diminishes with distance; the height of emission only influences air concentrations relatively close to the source.

A second misconception is that Scandinavia is, for some reason, being singled out as the recipient of excessive amounts of sulphur deposition compared with other parts of Europe. This is also untrue; a small mountainous area in Southern Norway has an enhanced deposition rate due solely to its large orographic rainfall, but it still receives much less sulphur than most of Britain and Central Europe.

A basic question to be asked is whether the evidence confirms that rainfall acidity is increasing, due to sulphate additions, at a rate that can be correlated with increases in SO_2 emissions, an association that is often claimed. Figure 10 shows the annual measurements of rainfall acidity at three monitoring sites in Norway which formed part of the European Air Chemistry Network (EACN). This kind of data is used to demonstrate a trend towards greater acidity over the past decade or so, and at first sight this seems very reasonable. However, the apparent upward trend is critically dependent on a rapid increase, confined to a year or two around 1965. Statistical analysis shows that the data are better explained as two periods of constant acidity with a step change upwards, in between. Other data have been published on the sulphate content of rain from a wider selection of EACN measuring sites

Deposition contours for UK emissions alone show a relatively small UK contribution to the sulphur deposited in Scandinavia. The calculations are based on a grid of 127-kilometre squares and automatically smooth out the local concentrations due to cities.

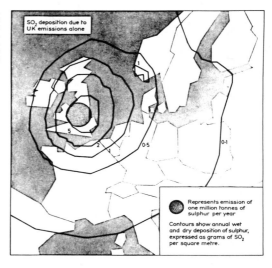

Figure 8

These computed contours for wet and dry deposition of sulphur (expressed as grams of sulphur dioxide per year) take into account all the European sources shown and assume average meteorological conditions. The locally enhanced deposition over southern Norway results from a special allowance for the high rainfall in this region.

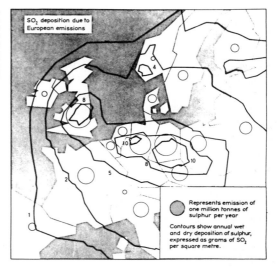

Figure 9

Emission from Coal-Burning Power Stations 61

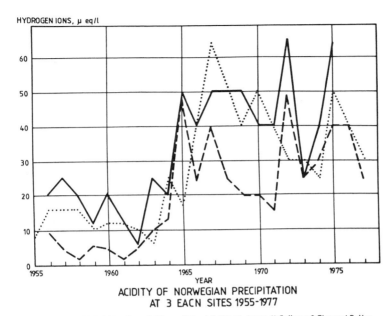

Figure 10

ACIDITY OF NORWEGIAN PRECIPITATION
AT 3 EACN SITES 1955-1977

From 'Acid Rain' by Gene E. Likens, Richard F. Wright, James N. Galloway & Thomas J. Butler.
Copyright © 1979 by Scientific American, Inc. All rights reserved.

in Scandinavia, over the same period. The same step change is evident, at the same period of time, and since then the sulphate content of their rain has steadily been decreasing, despite a 35% increase in European SO_2 emission. Explanations for a possible step change in conditions in the early 1960s have been sought in the meteorological records and this may be part of the answer. Another possibility, also discussed in technical publications, is that the earlier rainfall measurements by EACN were subject to sampling and analytical errors. It is known that the procedures for taking these measurements were altered in the early 1960s. If these, or other, factors can explain the presence of a genuine step change in the trend curves then the evidence for an upward trend that can be correlated with increases in SO_2 emission completely disappears. Similar conclusions can be drawn from published data on corresponding measurements of rainfall in North America.

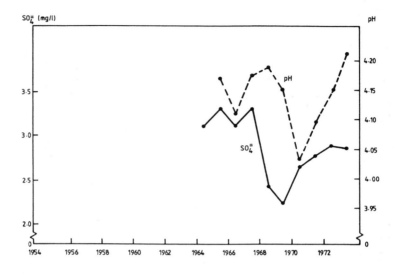

HUBBARD BROOK 1955-1974
ANNUAL WEIGHTED MEAN
CONCENTRATIONS IN PRECIPITATION
From Gene Likens et. al. "Hydrogen Ion Input to Hubbard Brook" 1976, Water, Air & Soil Pollution, 6, p. 436.
Copyright © 1976. Reprinted by permission of D. Reidel Publishing Company.

Figure 11

These data, shown in Figure 11, result from the Hubbard Brook studies in the USA and are the only consistent long-term figures from that country. Both the acidity (pH) figures and the sulphate in rain figures vary widely from one year to the next but show no discernible upward trend: indeed rather the opposite in the case of sulphate, as was found in Scandinavia over much the same time period. The single sulphate point for 1955 (which was derived from other sources and is rather speculative) reinforces the view that no trend is apparent, even over this longer time-scale. Furthermore, in the studies there was no significant relationship between the annual hydrogen ion input and the annual sulphate input to the catchment. No meaningful regression line could be fitted to the data in Figure 12, suggesting that variations in acidity were not related to sulphate input. A very different situation arose when hydrogen ion and nitrate ion were compared as shown in Figure 13. In this case, there was a highly significant correlation, suggesting that nitrates were responsible for much of the variation in acidity, in fact for 86% of the variation, compared with only 6% in the case of sulphate. This does not, of course, absolve combustion sources from responsibility but it does suggest rather strongly that limiting sulphur emission at source may not be a very effective answer to the problem. Surprisingly, data of this kind, published almost five years ago, have not in any way daunted the proponents of sulphur emission control programmes. However, it can only be concluded that firm evidence for an adverse long-term trend in rainfall acidity has yet to be presented. The extreme variations in acidity from year to year, seen in all the data, also emphasise the danger of comparing figures for single selected years - another favourite exercise in the popular articles. Almost anything can be proved this way by a suitable choice of dates.

The role of sulphates in rainfall acidity is also far from clear; their contribution is uncertain and trend data do not support a close correlation with SO_2 emission rates, either in Northern Europe or in the USA. Another area of legitimate doubt arises after the rain has fallen. It is frequently assumed that, in areas of granitic rock with thin soils, the rainfall runs off into water-courses with little chemical change, so that the rain acidity is reflected in the acidity of streams and lakes. This, again, is not born out by actual measurements.

In the area of the Tovdal River in Norway - one of the rivers said to be most affected by acid precipitation - a comparison of the ionic composition of rainwater and river water (Figure 14) shows some very significant differences. Space does not allow a detailed discussion, but the difference in hydrogen ion concentration is particularly important. Rainwater composition is, in fact, considerably modified by contact with tree canopies, with litter layers, with the

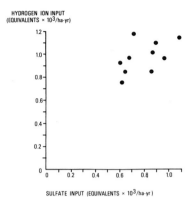

Figure 12

HUBBARD BROOK 1964 - 1973
RELATIONSHIP BETWEEN THE ANNUAL H⁺ION INPUT
AND THE ANNUAL SULFATE INPUT

From Gene Likens et al. "Hydrogen Ion Input to Hubbard Brook" 1976, Water, Air & Soil Pollution, 6, p.436.
Copyright © 1976. Reprinted by permission of D. Reidel Publishing Company.

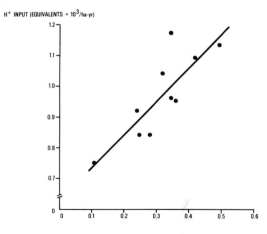

Figure 13

HUBBARD BROOK 1964-1973
RELATIONSHIP BETWEEN THE ANNUAL H⁺ION INPUT
AND THE ANNUAL NITRATE INPUT

From Gene Likens et al. "Hydrogen Ion Input to Hubbard Brook" 1976, Water, Air & Soil Pollution, 6, p.436.
Copyright © 1976. Reprinted by permission of D. Reidel Publishing Company.

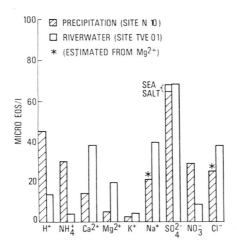

Figure 14

WEIGHTED AVERAGE COMPOSITIONS OF PRECIPITATATION AND RIVERWATER IN THE TOVDAL REGION OF S.W. NORWAY APRIL 1975 TO OCTOBER 1976

various soil horizons, and with the bedrock itself. Weak organic acids entering the streams from local biological sources can account for as much as half the total stream acidity and act as a buffer against strong mineral acids in rain. These local acidification processes add a further dimension to the problem of defining the contribution from distant combustion sources and estimating the benefit that would follow from limitation at source.

There are other factors that could be questioned but enough has been said to justify a cautious approach to the acid rain problem. It is an excellent example of the kind mentioned earlier, where controls applied remote from the point of impact may be inefficient and far from cost-effective. It is also a case where development of control policies is being hampered by a lack of scientific precision on the actual impacts being experienced, both in magnitude and in causation.

To embark on an SO_2 limitation policy in our current state of uncertainty would be a multi-million pound gamble against long odds of achieving some material benefit.

It is time now to turn from SO_2 and mention more briefly the other major pollutants from coal combustion in power stations, firstly nitrogen oxides. Natural sources of NO_x emission, mainly biological, probably account for the greater proportion of this gas in the atmosphere. Of the man-made sources, motor vehicles contribute the largest and the most rapidly growing part. NO_x from power generation accounts for one-quarter to one-third of the man-made emission.

The effects of NO_x emissions can be considered at different ranges of distance, as with SO_2. At short range the criterion is again the maximum concentration in the ambient air and its possible effects on public health. About 95% of the NO_x is emitted as nitric oxide (NO) and this oxidizes over several hours to nitrogen dioxide (NO_2). At the point of maximum concentration from a specific source, however, it is still mainly NO; which is less toxic than NO_2, which in turn is generally considered to be of similar toxicity to SO_2. It is apparent, therefore, that measures taken to control local SO_2 concentrations to safe levels will be more than adequate to control those of NO_x. A modern power station will contribute under 1 p.p.b. of NO_x to the annual average levels in its vicinity, and only a few p.p.b. to the daily averages. These contributions will nowhere approach recommended health limits.

At middle distances, NO_x can contribute to photochemical smog, if the appropriate conditions and precursors are present. These include the simultaneous presence of hydrocarbon emissions (other than methane), strong sunlight to power the photochemical reactions, and stable atmospheric conditions persisting long enough to allow a complex chain of reactions to occur. Neither of the latter conditions is particularly frequent in the UK but, on a few occasions, relatively high levels of ozone have been recorded which were probably the result of reactions of this kind. Although power stations emit NO_2, their emission of hydrocarbon is negligibly small, owing to high combustion efficiencies. In contrast, motor vehicles produce both NO_x and hydrocarbon in amounts that could lead to photochemical smog. An additional source of NO_x alone does not necessarily increase the risk of ozone production: it can in fact lead to something of an anomaly in respect of control policy.

For a given concentration of hydrocarbon in the atmosphere, increasing the NO_x concentration increases the ozone up to a certain point but then causes it to decrease again (Figure 15). This is because more of the ozone is taken up in oxidising NO to NO_2. The anomaly is that where NO_x concentrations are already high - that is, above the critical point for maximum ozone production - reducing the NO_x may make matters worse. Only when they are reduced below the critical point will ozone diminish. In contrast, reducing hydrocarbon concentrations will always lead to less ozone.

It has been suggested in the USA, in relation to control of motor vehicle emissions, that a policy directed principally at reducing hydrocarbon emission could be more effective than one aimed at reducing both hydrocarbon and NO_x together. Whatever the merits of this debate, it is clear that NO_x from all sources, including natural sources, will be

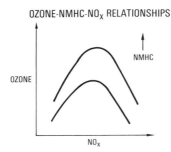

Figure 15

involved in photochemical smog, and that motor vehicles must be the first target of control, where this is found to be necessary. Although the NO_x contribution from power stations cannot be overlooked, it is not likely to be a critical factor.

The long-range contribution of NO_x to acid rain has already been indicated. As with SO_2, much more remains to be learned about the atmospheric chemistry of pollutants from combustion sources before useful progress can be made towards defining control policies. The CEGB contribution to the necessary research includes, among many other studies, a substantial programme of aircraft measurements over the North Sea, in collaboration with the US Electric Power Research Institute and the UK Meteorological Office.

We turn now to particulate emission: this makes an interesting contrast to SO_2 and NO_x in respect of control methods. Firstly, the impacts of solid emissions occur almost entirely in the short range - within a few km of the source. Secondly, methods for limiting dust emission at source have long been fully developed, are provided as a matter of course, and do not add more than 1% or so to the total cost of building and operating a power station. Thirdly, the scale of an uncontrolled emission would be such that dispersion alone could not possibly cope with the resulting problem.

The dust emission rate of 0.22 kg s^{-1} given in Figure 1 assumed a dust arrestor efficiency of 99.3%, which is the standard provided at all new coal-fired stations since the late 1950s. Without arrestment, the emission rate would be over 30 kg s^{-1} and, even with tall stack dispersion, this would lead to dust deposition rates of more than 700 mg m^{-2} day^{-1} a few kilometres away - at least five times the average rates for industrial cities. The actual deposition rates around power stations are monitored by standard gauges.

The trend of the average results around power stations, shown in Figure 16, is downwards over the years, at much the same rate as the trend for all gauges in the UK. The magnitude difference between the two curves is not significant;

DEPOSIT GAUGE RESULTS

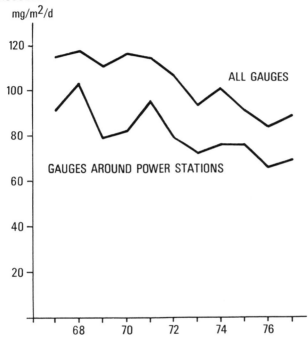

Figure 16 — YEAR ENDING MARCH 19

it merely reflects the fact that many new stations have been sited in rural or semi-rural areas, where the background rates are lower. The curves are for total insoluble deposit from all sources and the similarity of the trends indicates that the power station contribution to the lower curve is not very appreciable. Many other studies have confirmed the latter conclusion. It can be stated briefly, therefore, that if dust arrestor plant is provided to modern standards, and is adequately maintained and operated, the resulting dust deposition from power stations is not discernible against background rates and should not give rise to justifiable complaint.

In recent years, however, attention has been directed rather more to the elemental composition of the dust, rather than to its nuisance value, with some stress on the trace elements likely to be present - particularly the heavy metals. Whereas the coal before combustion has an elemental composition broadly similar to soils and crustal rocks - and hence similar to the natural dust content of the atmosphere - the combustion process acts to concentrate a number of elements into the ash and dust by a concentration factor of five or six. Beyond this, a number of the more volatile elements re-condense after combustion preferentially on to the finer particles - because of their greater specific surface area - enhancing the concentration of these elements by an even greater factor.

1. READILY INCORPORATED INTO SLAG		Al	Ba	Ca	Ce	Co	Eu
		Fe	Hf	K	La	Mg	Mn
		Nb	Rb	Si	Sm	Sr	Ta
		Th	Ti	Y	Zr		
	INTERMEDIATE	Co	Cu	Na	Ni	U	V
2. MORE READILY INCORPORATED INTO FLY – ASH		As	Cd	Ga	Mo	Pb	Sb
		Sn	Zn				
	INTERMEDIATE	Se					
3. REMAINING AS VAPOUR THROUGHOUT THE PLANT		Hg	Cl	Br			

Figure 17 **COAL TRACE ELEMENT BEHAVIOUR**

Figure 17 identifies the partitioning behaviour of a large number of elements, indicating those likely to be found in the furnace bottom ash, those in the fly ash, and those that are emitted still in vapour form. A large-scale survey is in progress to collect samples of fine suspended dust in the atmosphere in the vicinity of a large coal-fired station and to analyse the results for 36 different elements. The results to date have shown no perceptible effect of the station on the relative abundance of the different trace elements in the samples collected. All the levels measured are well below any maximum levels recommended in relation to human health.

The Effects of Changing Emissions on Biological Targets

M. W. Holdgate
DEPARTMENT OF THE ENVIRONMENT AND DEPARTMENT OF TRANSPORT, LONDON

Introduction

This paper is about how utilising more coal might affect living targets (man, animals, and plants) because of changes in the quantities of various pollutants emitted to the environment. It is not concerned with the direct environmental impact of mining, or the restoration of damaged landscapes following opencasting or the dumping of spoil. Nor is it concerned with the effects of emissions on non-living structures, through corrosion.

The paper commences with a brief account of the pollutants produced when coal is used, and an assessment of the relative contribution coal makes to these on a national and on a wider scale. The second stage in the analysis is to examine how these pollutants may affect man and other living targets (omitting the possible impact of carcinogens since these are treated fully in Professor Lawther's paper). Finally, an evaluation is attempted of how using more coal might change the scale and nature of these impacts.

Pollution caused by the Utilisation of Coal

The combustion of coal, colliery wastes, and products such as synthetic natural gas or liquid hydrocarbon fuels derived from coal is capable of producing:
    (a) Particulates and aerosols
    (b) Carbon monoxide and carbon dioxide
    (c) Sulphur oxides and their derivatives (especially sulphate)

---

* The views expressed in this paper are those of the author, and not necessarily those of the Departments of the Environment and Transport.

(d) Nitrogen oxides and the results of their interactions (especially with hydrocarbons, producing ozone and other oxidants)

(e) Hydrocarbons

The operations of mining can pollute the environment with airborne dusts, gaseous emissions from spoil heaps, and polluted run-off to watercourses. Power stations discharge heated cooling waters to watercourses (with or without enhanced salt concentrations). Finally, certain types of pollution control activity (notably flue gas desulphurisation) can generate wet sludges which themselves pose potential environmental problems.

Table 1 Relative contribution of coal and other fuels to USA domestic energy production, and associated air pollution (from OTA, 1979)

Fuels (Btu x 10^{15} yr^{-1})	1975	1984	1990
Oil	32.8	43.9	48.5
Natural gas	20.0	19.1	19.3
Coal	12.8	21.2	25.4
Nuclear	1.8	6.2	10.3
Hydro and geothermal	3.2	4.2	5.0
Total domestic	70.6	94.6	108.5

Pollutants (million tonnes)	1975	1985	1990
Sulphur oxides (all sources)	29.9	28.8	30.6
Sulphur oxides (coal burning utilities, industry, mine wastes)	21.3	–	19.6
Nitrogen oxides (all sources)	19.4	–	23.5
Nitrogen oxides (coal burning utilities, industry)	4.6	–	8.7
Particulates (all sources)	14.5	8.2	9.1

Table 1 (based on data in OTA, 1979[1]) summarises the contribution made by coal and other fuels to energy production in the United States in 1975 and its projected contribution in 1984/85 and 1990. It also indicates the national emission of sulphur oxide, nitrogen oxides, and particulates in the USA in 1975 and (estimated) in 1990 and the contribution coal is expected to make to these. Table 2 (based on DOE, 1978[2]) summarises the relative importance of coal, smokeless solid fuels derived from coal, and other fuels as sources of smoke and sulphur oxides in the United Kingdom between 1955 and 1975. The particular contribution of coal-burning power stations to these emissions in the UK is discussed in detail in Mr Clarke's paper.

Table 2 Relative importance of various fuels as sources of smoke and sulphur oxides in the UK

(million tonnes)

	1955	1960	1965	1970	1975
Smoke					
(a) from coal[1]	2.35	1.75	1.15	0.72	0.39
Sulphur dioxide[2]					
(a) from coal	4.32	3.93	3.84	3.34	2.79
(b) from other fuels	0.72	1.56	1.94	2.77	2.19

Notes 1. The combustion of solid 'smokeless' fuels, petroleum oils, and gas is relatively smokeless: in 1968 perhaps 5% of total national smoke emissions came from fuels other than coal.
2. Figures for 1970 and 1975 come direct from DOE (1978).[2] Those for 1955-65 inclusive come from Warren Spring (1972),[3] adjusted to make the categories more closely comparable with those used in the two later years.

Carbon dioxide is an inevitable product of the combustion of any carbonaceous fuel. Its concentrations in the atmosphere have been rising steadily from around 260-290 p.p.m. in the 19th century to around 330 p.p.m. in the early 1970s (SCOPE, 1977[4]). There is no evidence of direct effect on living targets: concern about these changes arises from possible modifications of temperature and rainfall, with especial consequences for agriculture. This subject is discussed fully in Sir John Mason's paper, and will therefore not be considered further here.

Over the world as a whole man emits far less particulate matter to the atmosphere than natural processes do. In the USA it is calculated that coal burning contributes 33% of man-made particulates, but 90% of the particulate matter emitted from 'stationary sources' burning fossil fuels.[1] In the UK smoke emissions have fallen steeply since 1960. In 1976, 0.33 million tonnes (Mt) out of a total of 0.37 Mt came from domestic fires and the remainder from industry (including collieries, public services, and agriculture). Fuel conversion industries (power stations, coke ovens, and smokeless fuel production plant) are relatively smokeless, and although smoke from vehicles can be significant locally it is negligible on the national scale.

Coal is one of the more important sources of atmospheric sulphur oxides. Weatherly (1977)[5] states that on average coal contains 1.2-1.5% S, whereas fuel oil has 2.5-3.0% and other petroleum 0.04-0.7%. Worldwide, man causes the emission of perhaps 65 Mt of sulphur annually in the form of SO_2, out of a total emission to air from all sources of around 144 Mt.[4] Coal is the source of over half the man-made SO_2 in the United States[1] and of 80% of man-made emissions from combustion in stationary sources (which together produce 70% of all man-made SO_2). It will be noted from Table 1 that coal burning in the USA is expected to double between 1975 and 1990, but emissions of sulphur oxides from this source are not expected to rise by more than 20% as a result of strict emission controls - a point returned to in the concluding section of this paper. In the UK coal was the source of nearly 60% of the total 1976 sulphur oxide emission of nearly 5 Mt.

Man-made oxides of nitrogen probably account for only about 10% of the total annual world production (about 50 Mt out of 500 Mt). United States figures suggest that coal burning is the source of about 24% of the NO_x made by man, but 49% of the NO_x from stationary sources. Coal has a high nitrogen content relative to other fuels, and modern combustion plant is designed to burn fuel-rich in the initial stages, releasing the nitrogen unoxidised before combustion is completed.

Finally, coal is a very minor source (under 2% in the USA) of those hydrocarbons whose interactions with nitrogen oxides are important in producing oxidant smog, and is also the source of various trace elements such as mercury, chlorine, fluorine, arsenic, cadmium, selenium, other metals, and radionuclides (Pennsylvania anthracite contains up to 4.7 p.p.m. thorium, and Gulf lignite 2.4 p.p.m. uranium).

Mining operations create pollution on a far more localised scale. In the USA the combustion of colliery spoil heaps and abandoned workings may contribute between 0.6 and 1.8% of total national emissions of carbon monoxide, sulphur oxides, hydrocarbons, nitrogen oxides, and fine particulates as well as some hydrogen sulphide and ammonia.[1] Acid mine drainage can be a serious local problem in the Appalachians, where the topography facilitates water flow through old workings and where the coal and bedrock are high in pyrite: about 6000 miles of watercourse and 15,000 acres of open water are affected (Bradshaw, 1973[6]). In England and Wales 346 separate emissions of mine water to non-tidal rivers were noted in 1975:[2] some of these caused water quality problems because of acidity, heavy loads of suspended solids, high concentrations of metals and other dissolved substances, or deoxygenation of the receiving water.

Water discharged to rivers and lakes after use in power station cooling systems has two potential ecological effects. First, the return of water used for direct cooling raises the temperature of the receiving waters, altering their nature as a habitat, and second, water returned from evaporative cooling systems, though creating less thermal pollution, may contain twice the initial salt concentrations. In the USA it is calculated that some 6 Mt a year of dissolved salts - 28% of all 'dissolved solids' in emissions to watercourses - comes from the cooling systems of utilities. In England and Wales in 1975 some 18 million m^3 of cooling water were discharged daily from power stations and other plants to non-tidal rivers, and another 40 million m^3 to tidal rivers, but 98% of the former and virtually 100% of the latter was recorded as 'satisfactory' in quality (i.e. it met the conditions laid down by the river authority).[2]

Flue gas desulphurisation often yields a calcium sulphate or magnesium sulphate sludge with a consistency of thin toothpaste. These sludges are generally treated by dewatering and can then be dumped on land under controlled conditions with only local ecological consequences.

The Impacts of Gaseous Pollutants on Living Targets
Effects on Man and Animals. 'Particulates and aerosols' released by coal burning contain many substances besides unburned carbon: notable among these are metals and metalloids (arsenic, chromium, beryllium, cadmium, mercury, selenium), radionuclides, organic compounds (some of which are alleged carcinogens), silica and other minerals. Sulphates and nitrates may occur in liquid droplets or in combination with the particles. The finer particulate matter (under 3 µm in diameter) is both most difficult to remove by filtration or precipitation and most liable to penetrate into the lung, transporting adsorbed materials with it.

Concern over the effects of these particulates and aerosols has been focused especially on the ability of some particles and acid droplets to cause pulmonary irritation and chronic obstructive and restrictive lung disease. Sulphur dioxide gas almost always acts on human respiratory systems in conjunction with particulates and aerosols although it is itself readily soluble and removed in the upper part of the airway to the lung, where it can cause irritation and increase the resistance of the airway to gas exchange.

The acute effects of particulates and SO_2 have been well demonstrated in a number of pollution incidents, especially the 1952/3 London smog (Royal College of Physicians, 1970[7]). It is well established that 24-hour concentrations of around 500 µg m^{-3} SO_2 together with 150-350 µg m^{-3} smoke particles are associated with various respiratory symptoms including increased asthma and with elevated hospital admission rates. The various standards established for urban air around the world generally prescribe 24-hour limits of between 150 and 350 µg m^{-3} SO_2. The World Health Organization has published a review of all existing health evidence in 'Environmental Health Criteria' (WHO 1979[8]). They conclude that harmful effects to human health might occur where 24-hour smoke and SO_2 concentrations were both above 250 µg m^{-3} and recommend a longer term guideline for these pollutants in the range 100-150 µg m^{-3}. The associated annual means are respectively 100 and 40-60 µg m^{-3}. A draft EEC Directive based on these recommendations was discussed on 17 December 1979. It contained proposed mandatory limit values based on the WHO health effect figures. Additionally it proposed that all member states endeavour to move towards the stricter WHO guideline figures.

Nitrogen oxide concentrations are high only near to sources of pollution because they are reactive substances, rapidly modified in the processes generating peroxyacetyl nitrates (PAN), ozone, and other oxidants. NO_x is readily soluble and enters the tissues in the upper part of the human airway, but healthy people are not particularly sensitive to it. Some (contested) experiments state that there is an immediate effect on bronchitics and asthmatics, leading to respiratory resistance, at concentrations of above 200 µg m^{-3} (which can be reached for short periods near coal-burning power stations). The WHO recommendation is for an upper 1-hour concentration of 320 µg m^{-3} for NO_2 (which is exceeded at peak in some cities).

Ozone acts on man by itself and in combination with other pollutants. Nose, throat, and eye irritation is said to occur at around 200-300 µg m^{-3}, and breathing difficulties in exercising adults and chest pains in children have been reported at the same levels. More severe symptoms are said to occur in hypersensitive people at about 500 µg m^{-3} and in healthy young adults at 750 µg m^{-3}. The data are, however, not clear-cut, and there are discrepancies between American and Japanese figures which may reflect different interactions between ozone and other air pollutants in Tokyo and Los Angeles. There is also some evidence of adaptation: Canadians recently arrived in Los Angeles appear less tolerant of oxidant levels there than native Californians. Overall epidemiological analyses suggest no evident relationship between short-term oxidant exposures and mortality rates.[1]

It has been alleged that sulphate aerosols have effects on

man that are distinct from those of sulphur dioxide and this has caused some concern in the USA, partly because whereas urban SO_2 concentrations have fallen those of sulphate aerosols (which are liable to long-range transport from power stations) have not. There are allegations of aggravated asthma and increased risk of chronic bronchitis at 24-hour sulphate aerosol concentrations as low as 10 µg m^{-3} (frequently encountered in summer in the North-Eastern USA). However, other work concluded that ammonium sulphate, zinc ammonium sulphate, zinc sulphate, and sodium sulphate have no effect at concentrations up to 1000 µg m^{-3} (Sackner et al. (1977)[9]) while WHO (1979)[8] reports no evidence that sulphate particulates have a harmful effect on man in short-term chemical experiments. Sulphuric acid aerosols are reported to have some effect on man and animals at concentrations above 750-1000 µg m^{-3}. The general balance of evidence suggests therefore that, although sulphuric acid aerosol alone or in combination with other pollutants can have health effects at levels higher than current ambient concentrations, there is no evidence that it, other sulphates, sulphur dioxide, or particulates are hazardous at the kinds of level generally encountered in urban air in North America or Europe.[1]

The general conclusion from direct observations, epidemiological analyses and laboratory studies is that air pollution is not killing people at present urban concentrations. This conclusion appears to conflict with reports of a recent epidemiological analysis by the Brookhaven Laboratory.[1] This study correlated human mortality rates in various urban communities with airborne pollution (presumably used as indicators of general air pollution). From the correlation was derived a coefficient of air pollution-induced mortality of 3.25 deaths per 100,000 people per year per 1 µg m^{-3} sulphate. Were this figure correct the implication would be that coal combustion today was responsible for about 48,000 deaths per annum in the USA (some 2.5% of total mortality), while the projected increased coal combustion could raise the annual air pollution-induced mortality by 7000 deaths a year in 1990.[1] However, there are many reasons for questioning this conclusion. The analysis appears to present a 'worst case', and a 'no effect' relationship is also possible within the likely confidence limits. Moreover, as OTA (1979)[1] comments; the correlation relates all deaths in each geographical area to air pollution defined from a single centrally sited monitoring station. The study may well have confounded the influence of air pollution with that of other variables like occupational exposure (inner urban areas may be the site of less healthy employment), smoking (which may be higher in urban populations, and could act synergistically with air pollution on chronic bronchitics)[7], and differences in population structure, due to the migration of healthy young people from the urban areas where employment is least available and social and environmental deprivation worst. The data need much fuller analysis, backed by clinical studies, before they can be used as a basis for policy. Nonetheless, the possibility of a significant 'urban effect' on mortality, related to air pollution, clearly needs further exploration.

Little information is available about the effects of these kinds of air pollution on farm or wild animals (in contrast to the substantial body of knowledge about the impact of airborne fluorine on cattle and sheep). Various laboratory experiments suggest that there may well be broad correspondence between impacts on man and impacts on other mammals. There is virtually no information about the effects of the air pollutants discussed above on invertebrates or on the animal component of wild ecosystems.

Effects on Plants. Sulphur dioxide, ozone, and nitrogen oxides all affect plants because they enter the leaves, mainly through the stomata, and affect palisade and mesophyll cells causing cell collapse, damage to the plastids, and irreversible loss of photosynthetic capacity (ARC, 1967,[10] Saunders, 1976,[11] Holdgate, 1979[12,13]). Sulphur dioxide also damages lichens by killing the algal cells on which they depend for photosynthesis. Particulate pollutants influence these processes by blocking stomata and by modifying the properties of plant surfaces which they contaminate. Most research has been done on the effects of SO_2 and O_3 and rather less is known about nitrogen oxides and sulphate aerosols.

Laboratory studies suggest that sulphur dioxide on its own impairs the growth of the sensitive S23 strain of ryegrass at concentrations around 50-100 µg m^{-3} (Bell and Mudd, 1975[14]), this plant being more readily affected in winter than in summer. Lichens are more sensitive, some species disappearing from areas with mean winter SO_2 concentrations above 30-50 µg m^{-3}. Many coniferous trees are also sensitive, and cannot be grown in urban centres or down-wind from some industrial areas. Other plants, including most crop species, are more robust, but damage to many can be expected at concentrations above 500 µg m^{-3}. There are considerable uncertainties, especially as most of the results have come from growth chamber studies with constant pollution levels, but the overall pattern can be summarised (T M Roberts, personal communication) as follows:

- 850-1400 µg m^{-3} for 1-3 h produces visible injury in many crop species.
- 200 - 400 µg m^{-3} for 1-3 months produces significant yield losses in the limited number of crop studies.
- 100 - 200 µg m^{-3} for several months has produced small yield losses in some studies but not all.
- 50 - 100 µg m^{-3} for several months can produce beneficial or detrimental effects.

<u>Ozone</u> is also well known to damage certain species and strains of plants at concentrations in the range 100-200 µg m^{-3}. Potato leaves have been damaged after 4 h at 200 µg m^{-3} and white pine (*Pinus strobus*) needles after the same period at 130 µg m^{-3}. Other species are more resistant, the pine *Pinus taeda* tolerating ozone at 100 µg m^{-3} for 18 weeks without visible injury (although with some chronic effects on leaf growth and photosynthesis) (references cited by Holdgate[12]). Ozone concentrations exceeding 100 µg m^{-3} for 4 h or more have been recorded in Southern Britain on a number of days in summer, and peak levels four times as great are found in many American cities. Plant damage from this cause is well known and widespread in parts of North America (in 1962-63 it was recorded over a zone 100 km north and south from Los Angeles), and can be expected also in Europe. There is no doubt that the UK itself is a major source of the ozone and oxidants

detected in rural areas of southern England (such as Harwell): it is also highly probable that it drifts northward from continental Europe under anticyclonic conditions, thereby providing another instance of trans-frontier pollution. Sensitivity to SO_2 and to O_3 is not only affected by genetic inter- and intra-specific differences but by the concentrations of other pollutants in the atmosphere. SO_2 and NO_2, or NO and NO_2 in combination have more effect on the rate of photosynthesis in the pea *Pisum sativum* and tomato respectively than either does alone (Figure 1). Ozone and SO_2 can also act together: one study demonstrated 25% and 58% damage to the leaves of two strains of tobacco at concentrations that were ineffective when the gases occurred singly (Menser and Heggestad, 1966[15]). Depending on target species, environmental conditions and concentrations it is possible to get an apparent synergistic, additive or antagonistic interaction between SO_2 and O_3 (Bell, quoted by Holdgate[12]).

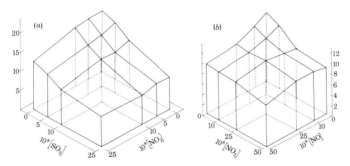

Note (a): Fumigation conditions: 1.20 air changes min^{-1}; temperature, 21 °C; light intensity, 73 J $m^{-2}s^{-1}$; water vapour pressure deficit 995 pa (from Bull and Mansfield, 1974); (b): From Capron and Mansfield, 1977.

Diagrams from Holdgate[12] reproduced by permission of the Royal Society.

Figure 1 (a): Effects of SO_2 and NO_2 pollution on the rate of net photosynthesis in pea (*Pisum sativum*); (b): Effects of NO and NO_2 pollution on the rate of net photosynthesis in tomato

As with effects on man, extrapolation from laboratory to field is not easy and there are doubts about the scale of agricultural and forest losses due to ambient air pollution. In 1952 Bleasdale (1959)[16] estimated losses of £2.6 million per annum to agriculture in South East Lancashire due to smoke and sulphur dioxide. The national cost of agricultural loss was set by the Beaver Committee (1954)[17] at £10 million (out of an estimated total air pollution damage cost of £250 M) while a later analysis (PAU, 1972[18]) suggested £39 million ± £11M. In the USA damage to crops was assessed at $100 million in 1968. These figures do not take account of the fact that yield losses can occur without visible injury: attempts to allow for this suggest a two- or three-fold increase in damage costs, but the data are very uncertain.

It is fair to add that not all the impacts of airborne sulphate and nitrate are damaging. Where soils are sulphur-deficient (as in some sandy districts of Eastern Britain) prolonged exposure to air containing around 50 µg SO_2 m^{-3} may be beneficial. Higher urban SO_2 levels are credited with preventing rose black spot, rose mildew, and maple leaf tar spot, all caused by sensitive fungi. Oats and wheat rusts and barley powdery mildew are also sensitive to low levels of pollutants. Nitrogen oxides washed onto the land as nitric acid or nitrates could also be beneficial: PAU[18] point out that annual nitrogen emissions in the USA amounted then to around 3 Mt as against 9 Mt added to land as nitrogenous fertiliser.

The impoverishment of woodland ecosystems by the reduction in abundance of corticolous lichens is well known (Ferry, Baddely and Hawkesworth, 1973[19]): the effect is evident over quite large areas of industrial Britain (Figure 2). It may be deduced that some conifers and crop plants may be reduced in abundance or performance in areas with winter SO_2 concentrations above 50 µg m^{-3}, while there may also be subtle ecological effects. But in recent years the most talked-about ecological impact of air pollution from fossil fuel combustion has been that allegedly due to 'acid rain'. This is predominantly caused by the wet deposition of SO_2 oxidised to sulphate and deposited

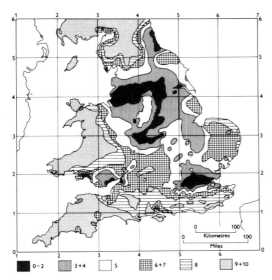

Note. Zone scale for the estimation of mean winter sulphur dioxide levels in England and Wales using corticolous lichens (adapted from D.L. Hawksworth and F. Rose, Nature (London), 1970, 227, 145-148).

Figure 2 Approximate limits of lichen zones in England and Wales

in rainful as dilute sulphuric acid (Smith and Hunt, 1979[20]). Some NO_2 may likewise rain out as dilute nitric acid. These acids have the potentiality to damage vegetation, possibly indirectly through changes in poorly buffered soils, and possible losses of forest productivity of around 10% over 30 years have been alleged (Sweden, 1971[21]). Other possible effects of acid rain on aquatic ecosystems are widely quoted. In the Adirondack Mountains of New York State some 51% of the lakes are said to have a pH under 5.0 and 9% of these now support no fish.[1] In Scandinavia lakes overlying granitic rocks and surrounded by acid podzols have little buffering capacity and are said to have shown a trend of increasing acidification ascribed to wet deposition of sulphate, with accompanying losses in trout and salmon fisheries.

The extent of these changes, and their detailed causative mechanism, are uncertain. Complex interactions between the deposited acids and the soil, affecting the leaching of other substances, are undoubtedly involved. Recent analyses[20] suggest that Britain was the source of about 30% of the total deposition of sulphur in Norway in 1972-74, and it has been commented that the 'tall stack' policy which has secured greatly improved dilution of our domestic SO_2 emissions from power stations and other major plants does nothing to abate this remote wet deposition. In the USA flue gas desulphurisation is used extensively and there may be pressures to adopt comparable policies for new plant in the UK (see next section of this paper). Whether this is justifiable depends crucially on the validity of the conclusions drawn about the significance of acid rain, and these are debated more fully in Mr Clarke's paper.

The Implications of Using More Coal

Simplistic analyses suggesting that doubling coal burning must inevitably raise atmospheric SO_2 and NO_x concentrations because coal contains more sulphur and nitrogen than do natural gas or most oils are clearly nonsense. American figures imply that doubling coal consumption need only raise SO_2 emissions by 20% (an increment of 3 Mt on 15 Mt). Moreover, if relative thermal values are taken into account coal contains less sulphur than many fuel oils do. The need is to examine for each pollutant how much more is likely to enter the atmosphere if use is raised by a specific margin and what pollution abatement techniques are available at what cost.

The United States forecasts in Table 1 are for an increase between 1975 and 1990 from about 590 Mt to 1175 Mt of coal used per annum - and since an increase in oil consumption is also forecast, none of this is offset by reductions in other fuels. Table 3 sets out the implied net change in air pollution assuming that this increase in coal use was accomplished without changes in abatement technology. The estimates for smoke and SO_2 in the UK assume that coal utilisation rises from its 1975 level of around 100 Mt to the 130, 160, or 200 Mt yr^{-1} suggested as low, probable, and high options by Ezra (1978)[22],

Table 3 Estimated total emissions of pollutants from coal combustion assuming no improved pollution abatement

		1975	1990
(a)	United States		
	Sulphur dioxide	21.3	42.3
	Nitrogen oxides	4.6	9.2
(b)	United Kingdom		
	Sulphur dioxide	2.79	4.46 { $^{+\,1.16}_{-\,0.83}$
	Smoke	0.3a	0.62 { $^{+\,0.15}_{-\,0.11}$

again without off-setting savings in other fuels, but that the relationship between use and emission remained as in 1975.

These extrapolations are self-evidently unrealistic. In 1956 when 217 Mt yr^{-1} were used in Britain, much of the coal was burned in old-fashioned, inefficient domestic grates producing much smoke. The future will not resemble that past - although the 1956 situation is helpful as a 'worst case' confirming that even were there reversion to those conditions severe pollution impacts would be localised (albeit unacceptable).

The US analyses[1] indicate that over the next 12 years many power stations are expected to come to the end of their working lives and that the extra 12.6 Mt yr^{-1} of coal to be used in 1990 will largely go into new plant. The extension of controls to existing plants and the adoption of more efficient combustion and filtration in new installations are expected to reduce <u>particulate emissions</u> from 14.5 Mt in 1975 to 8.2 Mt in 1985, with only a small rise to 9.1 Mt in 1990.

Desulphurisation, at present only operated in 15% of coal-fired utility boilers, is likely to become universal in new plant (it is worth noting that this may involve a capital expenditure between 1985 and 2000 of $20 billion and annual operating costs of $6 billion: equivalent to 14% of plant capital and 18% of plant operating expenditure). Flue gas desulphurisation can remove 89% of the SO_2 and sulphate from stack gases, although the lime/limestone scrubbing methods currently favoured produce a slurry waste whose disposal creates other environmental problems. It is clear, therefore, that in the USA SO_x emissions from major utilities need not increase even though coal burning rises, and is, indeed, unlikely to do so. One calculation suggests that total US <u>sulphur oxide</u> emissions are likely to rise only from 29.9 Mt in 1975 to 30.6 Mt in 1990.

In the UK, however, flue gas desulphurisation is not favoured. The substantial improvements in ambient SO_2 concentrations near the ground that have been achieved over the past 20 years have resulted from changes in domestic fuel use and better dispersal of power station and industrial emissions through the use of tall chimneys. A quick computation indicates that it would be both costly and difficult now to switch policies and bring about major changes in total emissions over any short period. In 1975 out of a total national emission of 5.3 Mt of SO_x only 2.15 Mt came from coal-burning power stations.[2] Assuming a total retro-fit of all this plant, and achievement of the best USA levels of abatement, some 1.9 Mt yr^{-1} might be removed, 35%. The American estimate is that the introduction of flue gas desulphurisation costs $100 per kW for new plant and substantially more for retro-fit. At present there is some 40,000 MW of coal-fired generating capacity in England and Wales, so to achieve this modest result would be likely to cost around £3000 M, while operating costs would be increased by some 18%.

The cost and inconvenience of flue gas desulphurisation is an incentive to develop new technologies. Two are under intensive study: pre-treatment of coal to remove sulphur before it enters the chamber, and fluidised bed combustion, which removes it in the combustion process if lime is added. Fluidised bed desulphurisation is more attractive for new plant because the combustion process is 10% more efficient in energy terms as well as allowing virtually complete sulphur removal, as a solid waste which is easier to dispose of in the environment than the slurries from wet scrubbing.

At present, <u>nitrogen oxide</u> emissions are kept down through design of boilers to expel as much as possible of the nitrogen in the coal prior to oxidation and, more importantly, to reduce the air flow through the combustion chamber, hence limiting the amount of atmospheric N_2 available. Fluidised bed combustion is very satisfactory in this respect. Combinations of various strategies have reduced NO_x production in large utility boilers by 40-50% and a new system being developed by the EPA in the USA offers prospects of 85% reduction.[1] Desulphurisation of fuels may also be beneficial in incidentally removing some nitrogen compounds. Various methods of flue gas scrubbing have also been explored in the USA and Japan (capital costs at $60-$90 kW^{-1} are comparable with those for desulphurisation). In the UK action to control NO_x other than incidentally through fluidised bed combustion and advances in boiler design seems unlikely: there are, however, uncertainties about its impact on vegetation which will need continuing research.

Taken together, these analyses indicate that it will be technologically possible to ensure that emissions to air can be held down despite major increases in the use of coal. Clearly the appropriate devices vary with the size of the coal-burning plant and between new and existing installations, and some technologies, such as stack gas desulphurisation and NO_x scrubbing, are costly and create residues that may in turn have unwelcome environmental impacts. It is no part of the present paper to prescribe technological solutions: that will be for control agencies, especially HM Alkali and Clean Air Inspectorate. The point of this review is to show that they exist, and hence that increased coal utilisation need not be accompanied by increased damage to biological targets - assuming that the necessary investment in abatement technology is accepted.

REFERENCES

[1] 'The Direct Use of Coal. Prospects and Problems of Production and Combustion', Report by the Office of Technology Assessment, US Government Printing Office, Washington, 1979.

[2] DOE, 'Digest of Environmental Pollution Statistics, No.1', HMSO, London, 1978.

[3] 'National Survey of Air Pollution 1961-71', Warren Spring Laboratory, HMSO, London, 1972.

[4] 'Environmental Issues', SCOPE Report No.10, ed. M.W. Holdgate and G.F. White, Wiley, London and New York, 1977.

[5] M.-L.P.M. Weatherly, 'Fuel Consumption and Smoke and Sulphur Dioxide Emissions in the United Kingdom up to 1976', Warren Spring Laboratory Report LR258(AP), 1977.

[6] A.D. Bradshaw, Ecological effects, in 'Fuel and the Environment', The Institute of Fuel, London, 1973.

[7] Royal College of Physicians, 'Air Pollution and Health', Pitman, London, 1970.

[8] Environmental Health Criteria: Sulphur oxides and Suspended Particulate Matters, World Health Organisation, Geneva, 1979.

[9] M.A. Sackner et al., Amer. Rev. Resp. Dis., 1977, $\underline{115}$, Pt.II,240.

[10] 'The Effects of Air Pollution on Plants and Soil', Agricultural Research Council, London, 1967.

[11] P.J.W. Saunders, 'The Estimation of Pollution Damage', University Press, Manchester, 1976.

[12] M.W. Holdgate, Targets of pollutants in the atmosphere, Phil. Trans. Roy. Soc., 1979, $\underline{A290}$, 591-605.

[13] M.W. Holdgate, 'A Perspective of Environmental Pollution', Cambridge University Press, 1979.

[14] J.N.B. Bell and C.H. Mudd, Sulphur dioxide resistance in plants: a case study of Lolium perenne, in 'Effects of Air Pollutants on Plants', Society for Experimental Biology Seminar Series No.1, ed. T.A. Mansfield, Cambridge University Press, 1976.

[15] H.A. Menser and H.E. Heggestad, Science, 1966, $\underline{153}$, 424-425.

[16] J.K.A. Bleasdale, The effects of air pollution on plant growth, in 'The Effects of Pollution on Living Material', Symposium of the Institute of Biology, No.8, ed. W.B. Yapp, Institute of Biology, London, 1959.

[17] Report of a Committee under the Chairmanship of Sir Hugh Beaver, Command 9322, HMSO, London, 1954.

[18] 'An Economic and Technical Appraisal of Air Pollution in the United Kingdom', Programmes Analysis Unit, Chilton, Didcot, HMSO, London, 1972.

[19] 'Air Pollution and Lichens', ed. B.W. Ferry, S. Baddely and D.C. Hawkesworth, Athlone Press, London, 1973.

[20] F.B. Smith and R.D. Hunt, The dispersion of sulphur pollutants over western Europe, Phil. Trans. Roy. Soc., 1979, A290, 523-542.

[21] 'Air Pollution across National Boundaries. Impact on the Environment of Sulphur in Air and Precipitation'. Sweden's case study to the United Nations Conference on the Human Environment, Royal Ministry of Foreign Affairs/Ministry of Agriculture, Stockholm, 1971.

[22] D. Ezra, 'Coal and Energy', E. Benn, London and Tonbridge, 1978.

Legislative Matters Relating to the Burning of Coal

D. H. Napier
DEPARTMENT OF CHEMICAL ENGINEERING, IMPERIAL COLLEGE OF SCIENCE AND TECHNOLOGY, LONDON

Introduction

In some respects the whole consideration of the title of this paper may be summarily dismissed in the terms that the law is the law and it has been set down for all to observe. It is the duty of the citizen to know the law as it impinges on him; even more so it is the duty of companies and corporations so to do. However true such statements may be they are less than constructive when a changing situation is envisaged or when a new situation may arise. This paper in the context of the conference is written with a novel circumstance in mind. This is the utilisation of more coal in a social and legal climate that has already adopted new and improving standards of pollution control as compared to those prevailing when larger tonnages of coal than those presently used were being consumed.

In attempting to consider such a situation, it must be born in mind that the technology of utilisation of the coal in the future will be different from that in the past. It is likely that some coal will be burned in fluidised beds, other coal will be used to produce liquid fuel, and perhaps yet further tonnages used for the production of methane. The combustion processes of previous decades brought with them problems that are unlikely to recur. However, the newer techniques for burning large tonnages of coal also carry environmental problems.

Further and major aspects of these considerations are the topics of interpretation of law and who, in fact, are the law makers. In no case is this seen more clearly than in the drafting of regulations that are made under the Health and

Safety at Work etc. Act 1974. Following the suggestion of the Robens Report the process of consultation has been built-in to law making. Operators and users (those words taken in the broadest sense) are involved in drafting, commenting upon, and modifying proposed regulations.

In whatever manner coal is burned there is a body of legislation presently in existence that is addressed to the activity. The problems that arise and to which regulations are applied may, however, vary according to the plant and process used. Methods of combustion that come immediately to mind are:
> pulverised fuel combustion
> automatic stoker systems
> fluidised combustion
> coal and oil mixtures
> cyclone combustor

The temperature regimes in these methods may vary as may the case of sulphur retention and hence the size of the sulphur problem; the problems of incomplete combustion vary according to the methods employed.

In considering environmental effects it may be concluded that burning coal pollutes the environment and burning more coal produces more offence to the environment. Such a statement may be an over-simplification, since more coal burnt in larger units offers a better opportunity for efficient combustion than for the same amount of coal burnt in many small units. The example *in extremis* is afforded by the comparison of heating by the open coal fire in one of the early domestic grates and steam raising or hot water generation in a modern boiler plant. Not only is there opportunity in the larger unit to control the combustion and increase its efficiency but also to clean the exhaust gases and, perhaps, to retain sulphur in the ash.

Another aspect of burning more coal is that even those materials that are produced at low concentration will in the light of increased quantities require full consideration in respect of both local pollution and ultimate dispersion. Thus the possibilities of producing photochemical smog may arise for consideration with a conceivable requirement for the monitoring of NO, NO_2, SO_2, SO_3, O_3 and unsaturated hydrocarbons.[1] It is pertinent to comment that the law as presently written could be applied, in general terms, to the control of such a situation.

The Range of Problems

In view of the coverage of the papers given at this conference only brief reference need be made to the hazards to which the law has addressed itself. In so doing the legislation must be flexible and all-enveloping and thereby retain the necessary degree of control without stifling endeavour.

Among the problems are the following:

(i) Products of combustion, release of:
- carbon monoxide
- carbon dioxide
- NO_x
- SO_x
- particulates
- trace elements

(ii) Self-heating in storage:
- products of combustion as in (i)
- odour
- risk

(iii) Solid disposal:
- colliery refuse
- ash, clinker
- sludge from coal cleaning

(iv) Liquid effluents:
- mine water
- coal washings

(v) Offence to the landscape:
- storage and handling
- chimneys
- cooling towers
- grime in the vicinity of large amounts of coal

(vi) Noise

(vii) Waste heat:
- heat pollution
- creation of microclimates

Some of these aspects are illustrated by referring to values given by Slater and Hull[2] for burning of bituminous coal in all of the electricity generating plant in the USA and shown in Table 1.

Table 1

		Cyclone	Dry bottom Pulverised	Wet bottom Pulverised	Spreader Stoker
Particulates		250	3500	500	76
SO_x	$10^3 t\ yr^{-1}$	1900	10^4	1900	99
NO_x		1900	2400	760	20
H/C		7.6	41	7.6	1.2
CO		26	140	26	2.6
Sb	$t\ yr^{-1}$	63	35	6.3	0.33
As		410	2200	410	21
Ba		270	1400	270	14
Be		30	160	30	1.5
Bi		13	67	13	0.69
B		660	3500	660	34
Br		760	4000	760	39
Cd		7.3	39	7.3	0.38
Cl		76×10^3	4×10^5	76×10^3	3900
Cr		180	930	180	9.3

Structure of the Law

Attention must be directed here to several types of legal constraint and these will be briefly described.

Contract Law. This is applicable to the working rather than to the general environment. It is concerned with local agreements and conditions of employment negotiated between employer and employee. Indirectly it becomes of importance in negotiating conditions for mining, in discussion of risks in loading and unloading and considering risks from novel plant, e.g. pressurised fluidised beds.

Common Law. Employers owe a duty of care towards employees, visitors, and third parties. In broad terms the common law duty spreads over activities in all walks of life. The keynote of the duty is 'reasonableness' and in the work place this entails:

>provision of a safe place of work
>provision of safe equipment
>organisation and maintenance of a safe system of work
>employment and training of competent workers

These matters are considered by the Courts against a background of tradition and previous rulings.

At a lower level in the legal hierarchy environmental issues are often raised in the context of abatement of a nuisance; such matters are usually dealt with in a court of summary jurisdiction.

Statutory Law. There is a sizeable body of legislation relating to the environment and therefore of concern in the context of utilising more coal, i.e. more burning coal-winning, handling, transportation, more products of combustion and coal processing, etc.

The statutes are available and are enforced as such. Some will have the very important undergirding of:
- regulations
- codes of practice (approved)
- notes of guidance

The requirements set down often relate to a minimum rather than to a high or even average standard of practice. Inevitably there are delays in the process of law-making and situations as originally considered and examined will continue to develop. So also will the standard of the technology in the broad terms of health and safety at work and in the environment. This problem is met to some degree by writing codes of practice.

In the present context the statutes are concerned with:
- plant construction
- process hazards
- service and construction hazards
- handling and disposal of effluents

General Legislation

In this section the discussion will centre around the impact of the Health and Safety at Work Act. The outstanding feature in the present context is that this piece of enabling legislation impinges on so many fronts.

The Health and Safety at Work etc. Act 1974. This Act is addressed to employers and relates as in S.2 to employees. Therein the general duties of employers are set down for ensuring, as far as is reasonably practicable, the health, safety, and welfare of employees. It is noteworthy that the elements of common law have been enshrined in statute.

However, the Act proceeds to set out duties of employers towards those who are not employees but [as in S.3(1)] who may be affected by the employer's activities. The duty extends to ensuring as far as is reasonably practicable that there is no exposure to risks to health or safety.

A further aspect of this section [S.3(3)] is also of interest. It refers to cases that are prescribed, where it shall be the duty of every employer and self-employed person 'in the prescribed manner to give to persons (not being his employees) who may be affected by the way in which he conducts his undertaking the prescribed information about such aspects of the manner in which he conducts his undertaking as might affect their health or safety'.

There is also a general duty set out in S.5 that is owed by persons in control of any premises prescribed in S.1(1)(d), i.e. where the control of the emission into the atmosphere of noxious or offensive substances from the premises is prescribed. The duty relates to:

(i) use of best practicable means for preventing the emission into the atmosphere and for rendering harmless and inoffensive such substances as may be emitted

(ii) adequate use and supervision of the plant provided, i.e. best practicable means

The reference to offensive and noxious substances is couched in broad terms and as in the reference [S.3(4)] to persons who have control of premises applies to those used for trade, business, or other undertaking (whether for profit or not).

The situation relating to the use of S.5 has changed since the conference took place. A Consultative Document[3] has been issued concerning amendments to the lists of scheduled works and noxious or offensive gases. The Draft Regulations will be made under Ss. 1(1)(d), 5(3), 15(1) and 3(c), 49(1), and 80(1) on the Health and Safety at Work etc. Act 1974. The proposals include the repeal of Ss. 6 and 7 of, and the First Schedule to, the Alkali and etc. Works Regulations 1906 and S.11(2) of the Clean Air Act 1964. It is noteworthy that among the proposed amendments and alterations to the list of scheduled works is a reference to solid smokeless fuel and, in the list of noxious or offensive substances, to carbon dioxide.

Reference must also be made to S.6 in that more coal burning will require more plant and those who design, manufacture, import, or supply an article for use at work bear a duty in relation to health and safety. These are:

(i) to ensure as far as is reasonably practicable that the design and construction is safe and without risk to health
(ii) to carry out or arrange for testing and examination
(iii) to provide adequate information

From this enabling legislation it is worth noting in the present context that the following are broadly mentioned:
(i) release of noxious and toxic materials
(ii) use of best practicable means
(iii) protection of health and safety as far as is reasonably practicable for both employees and those not employed

The legislation is not couched in definite and narrow terms but is clearly widely embracing. It is a matter for speculation as to whether the activities likely to be undertaken in a 'coalplex' will bring such an installation within the scope of the proposed regulations of major hazards.[4] Some of the activities, e.g. solvent extraction and fluidised combustion, are still being developed. The amount of energy within such plants and the pressure of operation have yet to be optimised.

Specific Legislation

In this section attention will be directed to some aspects of the manner in which the body of legislation is addressed to specific aspects of burning coal, which is presently the main method of utilisation. The law has been concerned with certain of these aspects over a long period, e.g. generation and release of smoke as illustrated in the list below:

Railway Clauses Consolidation Act 1845 (railway engines required to consume their own smoke)

Town Improvement Clauses Act 1847 (smoke from factories)

Smoke Abatement Acts 1853-6 (applicable to London Metropolitan Area)

Sanitary Act 1866 (sanitary authorities to be used against smoke nuisance)

Public Health (London) Act 1891

Public Health (Smoke Abatement) Act 1926 (amendments to the 1875 and 1891 Acts with extensions)

Public Health Act 1936 [included Public Health (Smoke Abatement) Act 1926 and addressed to trade effluents also]

Alkali Works Act 1863 and 1906

Alkali Act 1863. This was originally concerned with the 'more effectual condensation of muriatic acid gas in alkali works'. Through subsequent acts and regulations taken under them, this legislation is used in the control of many processes. The Act contains two limits set by Statute, those for hydrogen chloride and sulphuric acid, but for well-established process the Alkali Inspectorate require the level of discharge to be below presumptive limits.

The original Act of 1863 was temporary, but it was made permanent in 1868. Its scope was extended by amendments culminating in the Act of 1906. The Act of 1881 had added 'etc.' to the subjects controlled and in the Act of 1906 (S.27) the phrase 'best practicable means' was defined. A more recent definition of the phrase occurs in Part III of the Control of Pollution Act 1974 and is as follows:

> 'best practicable means, means reasonably practicable having regard among other things to local conditions and circumstances, to the current state of technical knowledge and to the financial implications'

It is noteworthy that this important principle is perpetuated in the Health and Safety at Work etc. Act 1974 [S.5(1)]. These Acts and the regulations taken under them have established several important principles that have been developed over many years, viz.

(i) the capacity of a local environment to absorb and deal with a certain amount of pollution without unacceptable deterioration

(ii) the technical feasibility of preventing the escape of pollutants and the cost of such methods to be taken into account

(iii) industry is an essential part of the social structure

(iv) some environmental disturbance is inevitable, but it must be reduced to a minimum

The Alkali Inspectorate is concerned with the Fuel Industry and a perusal of the Chief Inspector's report deals with this under the general heading of Registered Works. References are made to the problems of emission of sulphur oxides and acid soot (also to orange coloured spotting after the addition of dolomite to neutralise the acidity; the dolomite contains iron) and particulate fall-out.

Detailed investigations on both fall-out from and gas-washing of power station gases are also recorded. This is in keeping with the registration of Electricity Works in which solid or liquid fuel is burned to raise steam for generation of electricity for distribution to the general public or for purposes of public transport or in which boilers having an aggregate maximum continuous rating of not less than 204 tonnes h^{-1} of steam and normally fired by solid or liquid fuel are used to produce steam for generation of electricity for purposes other than those mentioned in the preceding section.

Control of Pollution Act. The Act deals with regulatory methods and general problems under four specified headings:
- (i) collection and disposal of waste
- (ii) pollution of water (discharge of trade effluents interference with flow so that pollution is caused or aggravated)
- (iii) noise (from plant or machinery)
- (iv) pollution of atmosphere

This important Act, now almost fully in force, endows wide powers of control and modifies and extends existing enactments. The list of repeals at the end of the Act is a guide to a large number of areas involved in this legislation. These range widely, e.g.

Salmon and Freshwater Fisheries (Scotland) Act 1862
Salmon and Freshwater Fisheries Act 1923
Burgh Police (Scotland) Act 1892 and 1903
Clean Air Act 1956
Criminal Justice Act 1967
Deposit of Poisonous Waste Act 1972 (completely repealed)
Water Act 1973

Not inconsiderable reference is made to the role of the Local Authorities in the control of pollution including the measurement and recording of emissions.

The Building Regulations 1972. These regulations include a section, M.2, on the prevention of smoke emission in the terms of control of appliances used for heating or cooking. Such appliances must not emit products of combustion into the atmosphere unless they are designed to burn fuel gas, coke, or anthracite.

The Clean Air Acts 1956 and 1968. This is the main enactment in force in the UK limiting the emission of smoke; a number of regulations have been made under these acts. Under the 1956 Act all new furnaces except domestic furnaces rated at more than 55,000 Btu h^{-1} are required to operate as far as practicable without emitting smoke. 'Dark' smoke is as dark or darker than Shade 2 of the B.S. Ringelman Chart. 'Black' smoke is as dark or darker than Shade 4. Emission of dusts and/or grit must be minimised and it is required that new plant be fitted with apparatus for grit and dust arrestment. In this context recommendations have also been made for grit emission from incinerators.

Regulations have been taken relating to the necessary periods when dark smoke is likely to be emitted (Table 2):

Table 2

Dark Smoke (Permitted Periods) Regulations 1958

Permitted emission of dark smoke (Ringelman 2) in any period of 8 h

No. of furnaces served by chimney	No soot blowing during period min	Soot blowing during period min
1	10	14
2	18	25
3	24	34
4 or more	29	41

Incorporated in the legislation is the concept of declaration of the whole district of a local authority or a part thereof as a smoke control area; the penalties for offence, conditions, revocation, and variations of orders are set down.

Provision was also made for the setting up of the Clean Air Council. This body, made up of those with special knowledge experience or responsibility in regard to prevention of pollution of the air, both reviews and advises on this topic.

Also of interest in burning more coal are the Smoke Control Area (Exempted Fireplaces) Orders 1970, 1971, and 1972. These give conditional exemption to fireplaces whose chimneys may emit smoke. In brief the fireplace, the stoker, and the fuel used must be the specified combination. Certain domestic fireplaces (six in number) are exempt from the Orders if the washed coal (Housewarm) is used in them.

The Clean Air (Emission of Grit and Dust from Furnaces) Regulations 1971 define grit as < 76 µm in diameter; fume is smaller than dust (1968 Act); dust remains undefined.

The Clean Air (Measurement of Grit and Dust) Regulation 1971 empowers the Local Authority to serve notice of measurement and to make or arrange for these measurements to be made. Examples of the limits set down are given for Schedule 1 and Schedule 2 appliances in Table 3.

Table 3
Schedule 1 Boiler and industrial heating appliances

Maximum continuous rating of steam per hour from and at 100C 10^3 lb	Quantity of grit or dust h^{-1}; furnaces burning solid fuel lb
1	1.33
5	6.67
10	10
50	37
100	66
200	122
475	250

These show an alleviation over liquid fuels; grit extends from 0.28 to 57 µm.

Schedule 2 Indirect heating appliances and furnaces where combustion gas is in direct contact with material to be heated

Heat input h^{-1} MBtu	Quantity of grit and dust lb h^{-1} Furnaces burning solid fuel
1.25	1.1
5	4.3
10	7.6
50	14.1
100	35
595	250

It is appropriate again to direct attention to the fact that this large and important body of legislation on Clean Air has been produced during a period of declining coal usage. Further, this legislation has been developed while the open coal-fire, a major source of urban pollution, has declined in use and where used it has often benefited from considerable re-design.

Housing Act 1964 (S.95). The detail of this section is, in the present context, subsidiary in importance to the main subject. The Clean Air Act 1956 Ss. 12(1) and 13(1)(b) and (c) refer to contributions to the cost of making adaptations to fireplaces in dwellings so as to avoid the emission of smoke. The provisions were later amended in the Housing Act. This establishes the concept of community responsibility in these matters and sets a rating on their importance. Extension of the principle may be required if coal is burned in larger tonnage than at present in domestic premises.

Conclusion

In this paper attention has been drawn to some aspects of legislation relating to the increased utilisation of coal by both traditional and novel methods. It has also been shown that a comprehensive body of legislation is already in existence. It becomes clear that, should developments require it, further regulations can be made within the existing framework of statutes.

The present regulations are such that even if much more coal is burned there are powers available to prevent the disastrous situations that frequently occurred earlier in the century and reached a peak in the infamous London Smog of December 1952.

Passing reference has been made to several problems, e.g. NO_x, trace elements, heat pollution, novel appliances, and plants. These problems may still require further attention, as also does 'acid rain', but scientists and engineers continue to address themselves to these matters. The necessity for technology to tackle these problems and remain ahead of the demands of legislation has been recognised. It is in harmony with the spirit of both the Report of the Robens Committee[5] and of the Health and Safety at Work etc. Act 1974.

The World Health Organisation[6] has recommended long-term goals for levels of pollution. These are as shown in Table 4.

Table 4

Pollutant	Limit
Sulphur dioxide	Annual mean 60 µg m^{-3}; 98% of observations to be below 200 µg m^{-3}
Suspended particulates, i.e. smoke	Annual mean 40 µg m^{-3}; 98% of observations to be below 120 µg m^{-3}
Carbon monoxide (non-dispersive infra-red)	8 h average 10 µg m^{-3} 1 h maximum 40 µg m^{-3}
Photochemical oxidants expressed as O_3	8 h average 60 µg m^{-3} 1 h maximum 120 µg m^{-3}

Additions may be made to this list, e.g. metals, NO_x, hydrocarbons.[7] The future utilisation of coal must be so managed and developed that the above and any other problems that arise are tackled with foresight. The solution of the problems must be designed and developed by those who are closest to the action and on whom is placed this continuing duty in law.

REFERENCES

[1] S. Muthukrishman and L.R. Peters, A.I.Ch.E. Symposium Series No. 65, **73**, 1977, p.43.

[2] S.M. Slater and R.R. Hill, *ibid.*, p.291.

[3] Health and Safety Commission Consultative Document, 'Proposals for amendments to the lists of scheduled works and noxious or offensive gases', 1980.

[4] Health and Safety Commission Consultative Document, 'Hazardous Installations (Notification and Survey) Regulations', 1978.

[5] Safety and Health at Work, Report of the Committee 1970-72, Chairman Lord Robens, Cmnd. 5034 (1972) HMSO.

[6] 'Air quality criteria and guide for urban air pollution', WHO Technical Report Ser. No. 506, Geneva, 1972.

[7] 'Monitoring of the Environment in the UK', Pollution Paper No. 1, D.o.E. Central Unit on Pollution, HMSO, 1974.

Discussion on Session II

B Lees: Mr Clarke has illustrated very clearly that
pulverised fuel fired boilers fitted with electrostatic
precipitators operating at 99.3% efficiency emit 0.1% of the
fuel consumed as particulates. Oil fired boilers operating
without grit arrestors achieve similar results. As the
calorific value of coal is only about half that of fuel oil,
transfer of electricity generation from fuel oil to coal will
double the particulate emission assuming that the electrostatic precipitators operate with at least 99.3% efficiency.
A paper by Lees and Morley (J. Inst. F., 1960, 33, 90)
described simple equipment designed to measure daily on a
routine basis the emissions of dust and grits from coal fired
by single or multi-point sampling. Daily measurement is
necessary if optimum efficiency is to be maintained. Are all
modern CEGB pulverised fuel fired boilers fitted with routine
daily dust and grit sampling equipment?

A J Clarke: Whilst the calorific value of fuel oil is nearly
double that of coal, the volumes of flue gas produced for a
given heat output are similar. As the dust burden limitation
is the same for both fuels, the total dust output does not
vary substantially on changing from oil to coal. Not all
coal-fired boilers are yet fitted with continuous dust
monitors. The problems are well known; duct sizes are very
large, the dust distribution within them is not homogeneous,
and modern precipitators are divided into several parallel
paths and series zones. All this makes reliable monitoring
very difficult.

P A H Saunders (UKAEA, Harwell): Professor Lawther showed
this morning that the public health impact of, for example,

benzpyrene, was negligible compared with the effects of smoking. In the nuclear industry public pressure is leading to the spending of large sums of money for the reduction of risks that have been demonstrated to be extremely small. Does Dr Holdgate see the same pressures emerging in connection with carcinogens from coal burning?

M W Holdgate: The public perception of risk will not necessarily coincide with the results of dispassionate scientific analysis! But scientists have a duty to keep the community informed and so influence this public perception. Scientists must expect, however, to be called upon to do what is possible to allay public concern about problems they may not themselves regard as of high priority.

Sir John Mason: We have to accept the public's view of risk, even though the responsibility of the scientist is paramount. Sometimes too much is claimed too early: e.g. with Concorde and its effect on ozone, early statements were wrong and opinions have changed since. Similarly, the effects of chemicals on climate have been exaggerated in newspaper circles. How do we educate the public?

A J Clarke: We should also question some of the conclusions reached on the effects of acid rain.

P Mason (Institution of Structural Engineers): Environmental legislation will surely become part of the EEC's overall pattern. Will not this therefore lead to new standards to which we will have to subscribe, whether better or worse than our own?

D H Napier: This is a real danger. It calls for resistance to limits and test methods put forward with inadequate experimental and experential evidence. It also calls for us to take the initiative by proposing good and well-founded standards for the EEC to adopt.

E S Rubin (Cavendish Laboratory, Univ. of Cambridge): Is not a similar concept inherent in some of the US legislation, which has established standards of air quality? Is there any movement in the UK towards the adoption of such standards?

D H Napier: Yes, the situation in the US is, in some respects similar. Hitherto the adoption of limits on other than the basis of technological feasibility and pragmatism has been resisted in the UK.

G V Day (UKAEA): You say that Section 6 of the Health and Safety at Work Act calls for plant which offers no risk (as far as is reasonably practicable) to those not employed (the public). The Act appears to be very open-ended and presumably will rely on future case law. On whom is the onus of proof for injury, for example, in the case of sulphate particulate pollution of the atmosphere? Future precedent could lead to very onerous restrictions on flue gas emissions if present research shows sulphate particulates to be harmful.

D H Napier: The situation will become clearer from successive Court rulings in these matters. The corpus of case law on S.6 is as yet small; the phrase 'reasonably practicable' remains to be more fully explored. However, if an aggrieved party can establish injury there appears to be an absolute duty established in S.6 in respect of the effects on health and safety of items of plant and equipment.

T H Kindersley: We must remember that EEC regulations are not conjured up from nowhere; they do, in fact, come from some other European country.

A Keddie (Warren Spring Laboratory): Reference has been made to possible EEC Directives which would set air quality standards or guidelines. Dr Napier has criticised this approach to pollution control on grounds of impracticalities and that Threshold Limit Values (TLV's) are often used, without any justification, in deriving the numerical values of standards. I am sure that many participants are aware that negotiations on a Health Protection Standards Directive for smoke and sulphur dioxide are currently in progress. I will not comment on the pros and cons of that Directive but I would emphasise that the health evidence used as a basis for the figures in that Directive has been obtained from epidemiological studies (mainly by Professor Lawther's team in London) and has been agreed within the World Health Organization; there has been no arbitrary manipulation of TLV's.

D H Napier: This appears as a good illustration of setting standards where the back-up work is being done and so to place the standard on a good foundation.

E Raask (CEGB): Particulate emission now and 25 years ago has been discussed by Mr Clarke and by other speakers. The nature of solids emitted from coal-fired systems has markedly changed during this period. Domestic fires and small industrial boilers with a short flame path produce a great deal of unburnt coke and soot particles. The air-borne carbonaceous particles have a large specific surface area (surface to mass ratio) or about 20,000 to 50,000 m^2 kg^{-1}, and thus it is an ideal material for catalytic conversion of SO_2 to sulphuric acid and for transporting the acid into the respiratory system on inhalation. It is therefore inevitable that a combination of a high concentration of SO_2 and of carbonaceous solids as in previously experienced acidic smogs was particularly injurious to health.

By contrast, the particulate material which escapes the electrical precipitators of pulverised coal-fired boilers consists largely of spherical, non-porous particles of silicate glass with only 2-5% by weight of coke residue particles. The escaping ash has a comparatively low specific surface area, around 2000 m^2 kg^{-1} and the siliceous material is not acidic.

In some respects the flame-heated ash is biologically less reactive than natural, unheated siliceous dust. The respirable quartz particles in mine and quarry dusts can cause silicosis, whereas the quartz particles in pulverised coal are rendered inactive during their passage through the high-temperature flame. This change takes place as a result of vitrification and contamination of the surface layer of quartz by alkali-metal, iron, and aluminium compounds in the flame.

Session III

Chairman's Opening Remarks

G. S. Hislop
CHAIRMAN, COUNCIL OF ENGINEERING INSTITUTIONS

Mr. T.H. Kindersley, MA, FICE, FIMechE, FIGasE
Managing Director, Business Development Division, Babcock Contractors Ltd.

Mr Kindersley obtained a Mechanical Science Degree at Cambridge University and since then he has been a Resident Engineer on power station construction with Sir Alexander Gibb and Partners and has been in iron ore mining as Personal Assistant and then Deputy to the Chief Engineer at the Sierra Leone Development Company, Chief Engineer and then General Manager responsible for process machinery business in the Engineering Products Division of Allis-Chalmers (Great Britain) Ltd, and Director of Engineering of Woodall-Duckham Group Ltd, where he started the natural gas conversion division, Gascol Conversions Ltd.

As the Managing Director of the Business Development Division of Babcock Contractors Ltd, Mr Kindersley is responsible for marketing services, press and publicity, territorial development, business development, and strategic planning. He is Chairman of Babcock Jenkins Limited (bulk handling, stacker cranes, conveying systems, belt conveyor pulleys). Mr Kindersley is a member of the British Materials Handling Board.

Dr. P.C. Finlayson, BSc, MInstF, CEng
Deputy Managing Director, Coal Processing Consultants Limited

Dr Finlayson graduated from Leeds University with a doctorate in coal science. He worked for a number of years on energy utilisation in the Iron and Steel Industry before joining the

Babcock & Wilcox Group in 1965. While working for Babcock he was in charge of the company's development activities in fluidised-bed combustion. Since 1975, he has been Deputy Managing Director of Coal Processing Consultants Limited, a consultancy company set up by the National Coal Board and Babcock International Limited, to provide advice and assistance worldwide on the utilisation of coal.

Mr. F.E. Dean, BSc, CEng, FIGasE, FInstF, FInstPet
Chief Environmental Planning Officer, British Gas Corporation (Production and Supply Division)

Mr Dean has been at the Headquarters of the British Gas Corporation since 1962. Formerly he served as Principal Scientific Officer, H.M. Overseas Civil Services, Nairobi 1956-1962; Engineer, National Industrial Fuel Efficiency Services, Leeds and Cambridge 1954-1956; Engineer, Ministry of Fuel and Power, Fuel Efficiency branch, London and Leeds 1944-1954.

He has been a principal speaker at numerous national and international conferences on Environmental Impact Analysis and related subjects and has recently participated in the ECE Symposium at Villach, Austria, and the EEC Symposium in Brussels as an invited member of the UK delegations. Other lectures include those for the Society of Chemical Industry RECLAN Conference on Contaminated Land and for the British Ecological Society. He is currently a member of the CBI Environmental and Technical Legislation Committee.

Dr. B.B. Goalby, MA, BSc, CChem, FRIC, CEng, MIGasE
Senior Environmental Planning Officer, British Gas Corporation

Dr Goalby has been with British Gas Corporation since July 1977. He was formerly Director of the British Launderers' Research Association, Research Director and Head of Technical Services of Thomas De La Rue International Ltd.

His papers with the gas industry are concerned with pollution control, development of Contaminated Land, SNG and other site developments and other environmental matters.

He has been a speaker at various national and international conferences and is currently active with committees of the Society of Chemical Industry.

Dr. J. Gibson, FRIC, FEng
Member for Science, National Coal Board

Dr Gibson graduated in chemistry and engaged in research into coal constitution and carbonisation, gaining his MSc and PhD degrees.

He is a full-time member of the National Coal Board, with special responsibility for Science. Previously he was Director of the Coal Research Establishment of the NCB and Director of Coal Utilisation Research. He is a Member of the Advisory Council on Research and Development for Fuel and Power and a Director of several companies associated with the NCB: NCB (Coal Products), Nypro, Stavely Chemicals, Coal Processing Consultants, Consultants, etc. He is a member of the Fellowship of Engineering, Hon FIChemE, and was President of the Institute of Energy in 1975/76.

Mr. D.W. Gill, BSc, CEng, CChem, FInstE
Head of Pollution Section, Coal Research Establishment NCB

Mr Gill graduated with honours in chemistry (London University) in 1950 and worked on process development with ICI Ltd, Dyestuffs Division, Huddersfield from 1950 to 1955, when he joined the British Coal Utilisation Research Association's Leatherhead Laboratories.

He has had experience with BCURA in coal combustion and gasification and is joint author of a monograph on the combustion of pulverised coal (1967). He has been associated with fluidised combustion of coal since the late sixties, and was Superintendent of the Combustion Technology Department when BCURA ceased to function as an independent body in 1971. Since then he has worked on pollution problems and environmental control for the NCB at the Coal Research Establishment, Stoke Orchard.

New Technology in Coal Combustion

T. H. Kindersley
BABCOCK CONTRACTORS LTD., LONDON
P. C. Finlayson
COAL PROCESSING CONSULTANTS LTD., CRAWLEY, SUSSEX

Introduction

The in depth papers presented on the first day of the conference had, to an extent, one common theme, namely the extent to which the products of combustion of coal may or may not be harmful. The significant exception is the heat which is released from the combustion process and which an energy-hungry society demands in ever increasing quantities. The word 'demand' is used very deliberately because all those who are working in the energy field are under considerable pressure to find the ultimate in 'convenience energy' and the social environment appears to be becoming increasingly hostile.

Before coming to the more conventional elements of this paper, it is appropriate to comment that the environment, which is most fundamental to this conference, is not just the physical environment, with its excesses of oxides of carbon, sulphur, or nitrogen, but also the social environment. This social environment is the one to which scientists and engineers must devote at least as much attention as they do to technology. All human beings are in fact combustion units and consume significant quantities of food daily for no other purpose than to keep themselves warm and to provide themselves with energy which, through biological processes, emerges to a greater or lesser extent in the form of muscle power. It is outside the authors' competence to evaluate the human frame as a combustion unit, and it would be unfair to judge on twentieth century productivity and waistlines.

It would be interesting to hear how much of the increase in carbon dioxide is arising from an increase in the population and the destruction of those natural systems that nature devised for keeping the carbon dioxide in our atmosphere in balance, such as the Amazonian forests. This same society is, however, vociferous in its demand for more energy, for more convenient energy, for less politically vulnerable energy, and, particularly as far as this conference is concerned, for clean energy, which we are endeavouring to produce with no side-effects.

It is hoped that nobody will feel that it is inappropriate in a paper of this type to include these remarks, because it is strongly felt that our single-mindedness in trying to find economically and socially acceptable solutions to the needs of our fellow citizens will be much facilitated if we can not only have the benefit of their constructive criticism, but also receive some encouragement when there is a degree of success. A lead was given yesterday by Mr. Clarke in the matter of communication and the need for more of it. Perhaps more should be done to encourage our talented young men and women from the many and varied training grounds available, be they academic or industrial, to join with us in solving the energy problems rather than have them join the ever growing army of expert witnesses for the prosecution, if they might without offence be described as such. If the array of talent, both legal and technical, on both sides of the room, at some of the public enquiries about some of the major investments which will be the source of energy for younger generations following us, was all ranged on the side of doing what has to be done more quickly and at lower social cost, perhaps we would begin to move up the European or World league table for productivity and economic growth. Perhaps it is fortunate that Britain's oil was found offshore and not on land.

It is felt that, although it might be construed as departing from the strict definition of the conference, one might well be thought to have one's priorities wrong if no consideration was given to the environmental consequences of not burning more coal. Surely our prime responsibility is to ensure that there is enough energy when it is wanted and not ideal energy when it is too late to be of use. One need only compare the time required to plan and build a new coal-fired power station anywhere in the world and its related coal mine with the period since the Middle Eastern energy scene started a change which still has much potential for further development.

New Developments in Burning Coal

It is clear that during the period between now and the end of the century so called clean fuels are going to become increasingly short in supply. Therefore, we are going to be forced to use coal as a primary energy source. It will, therefore, become vitally important for us to make the best possible use of new developments which will allow us to burn coal cleanly and efficiently. These new developments fall conveniently into three categories:

(1) Conventional combustion followed by clean-up of the stack emissions
(2) New technology, such as fluidised bed combustion, which has the inherent capability to burn cleanly
(3) Conversion of the coal into a clean fuel

Later papers in the conference will be dealing with the conversion of coal into gaseous, liquid, and solvent fuels. Therefore, it is intended that comments be limited in this paper to the combustion of coal.

Pulverised Fuel Firing and Stack Gas Scrubbing

Pulverised fuel firing, the combustion of powdered coal, has been the most commonly used coal firing system for over half a century, particularly in the utility industry. This combustion system potentially leads to the emission of particulates and oxides of nitrogen and sulphur. However, techniques for the removal of particulates using scrubbers, bag filters, or electrostatic precipitators are now regarded as well established, and smoky chimneys are a thing of the past.

In such countries as Britain where low-sulphur coals are available, the CEGB has adopted a tall stack policy consistent with acceptable distribution of the small traces of sulphur dioxide in the products of combustion. However, in countries such as America, facing problems of using high-sulphur coals, there is growing use of stack gas scrubbing systems in spite of their cost.

There is a wide variety of stack gas scrubbing systems in commercial use. The most commonly used processes involve a 'throwaway' sulphur absorbent which may be in the form of lime, limestone, alkaline fly ash, or wet soda. The main disadvantage of these processes is that the sulphur is usually recovered as a calcium sulphate slurry, which poses disposal problems.

In view of these disposal problems, techniques which allow the recovery of the sulphur from the stack gases in the form of sulphuric acid or elemental sulphur are bound to find growing acceptance. These processes involve scrubbing the gas with magnesium oxide or sodium sulphite. On heating, the sulphur is released from the scrubbing solution as sulphur dioxide, whilst the magnesium oxide or sodium sulphite is regenerated for further use.

Installation by Utility companies in the United States of these processes is increasing rapidly from only a handful in 1972 to 42 operational plants in 1978 with an equivalent electrical capacity of 14,500 MW. By 1986 there are expected to be 135 operating systems with an equivalent electrical capacity of 53,000 MW. Of the present operational capacity approximately 93% is based on the throwaway lime/limestone processes and 3% on the wet soda process. The two recovery systems, which are more recent developments, account for only 4% of the scrubbing systems.

Fluidised Bed Combustion

Although one approach is to produce the sulphur dioxide from coal and then to remove it from the products of combustion by stack gas scrubbing, a much more satisfactory approach is to capture the sulphur dioxide as soon as it is formed. Fluidised bed combustion provides a method of achieving this objective.

A fluidised bed is formed when a bed of finely divided particles is subjected to an upward, low velocity air stream (Figure 1). At very low velocity the air simply passes through the bed of material. When the velocity is increased, the particles become turbulently suspended and the bed resembles a

New Technology in Coal Combustion 115

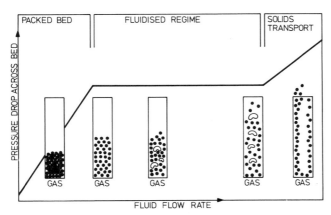

Figure 1 Fluidised bed behaviour with increasing gas velocity

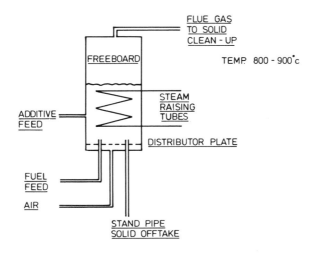

Figure 2 Atmospheric fluidised bed

bubbling liquid. At that stage it is said to be a fluidised bed. If the velocity is further increased the particles become entrained in the air stream.

Figure 2 shows the main features of an atmospheric fluidised bed combustor as applied to boiler firing. The bed is formed of sand, crushed refractory, limestone, or coal ash. The bed is heated up by means of burners which are directed into the surface of the bed. When the ignition temperature is reached, usually around 600 °C, the introduction of crushed coal at the base of the bed can begin. The fuel is burnt rapidly in the fluidising air and there is only a fraction of 1% by weight of combustible matter in the bed at any time. Heat absorption surfaces are immersed in the bed and are matched to the heat input so that the bed is maintained at a temperature of around 800-900 °C, which is well below the fusion temperature of most coal ash.

Development work in Britain on fluidised combustion started during the early 1960s. The National Coal Board in collaboration with the Central Electricity Generating Board investigated the combustion of coal whilst British Petroleum pioneered the application of fluidised beds for liquid fuel firing. By the mid-1970s these organisations had conducted some 20,000 h of rig tests. Babcock Power then became involved in the development programme and the decision was made to prove the technology by converting a 45,000 lb h^{-1} stoker-fired boiler to fluidised bed firing. The conversion was completed in April 1975 and this boiler at their works in Scotland has been operating successfully since that time. Figure 3 shows a photograph of this boiler at Renfrew with its associated control panel.

Figure 4 shows a sectional elevation through the boiler at Renfrew. All the main features can be seen including the light-up burners above the bed, the coal feed system, the ash discharge pipe, and the boiler heating surfaces both submerged in the bed and also in the freeboard. Since this is an experimental boiler, there is only a very simple dust collection system. For a commercial unit a bag filter or electrostatic precipitator would, undoubtedly, be required.

Figure 3   Photograph of Renfrew boiler

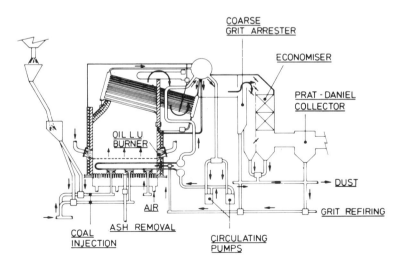

Figure 4   Schematic arrangement of Babcock cross-type boiler converted to fluidised bed firing

The experience gained in the operation of this boiler gives the confidence to comment on the advantages of fluidised combustion. It is regarded as being the most important advance in combustion technology, particularly from the environment point of view, for more than half a century.

The Advantages of Fluidised Combustion

Firstly we consider the control of sulphur dioxide emission. This is achieved through the introduction of limestone or dolomite in the bed. Crushed stone is continuously fed into the bed in order to maintain its sulphur retention activity. In Figure 5 it can be seen that the introduction of calcium oxide can lead to retention of over 90% of the sulphur in the bed. Thus it is possible to fire even very high-sulphur coals without recourse to stack gas scrubbing. In this figure it can be seen how the sulphur retention depends on the calcium to sulphur ratio, and also on the type of limestone which is used. The behaviour of limestone is influenced by both its chemical and its physical properties. The National Coal Board has therefore developed test equipment which can be used to predict the behaviour of a limestone or dolomite before it is used in a fluidised bed fired boiler.

The second important feature from an environmental point of view is the suppression of oxides of nitrogen. The combustion temperature in the fluidised bed combustor at around 800-900 °C is of course much lower than for a conventional stoker or a pulverised fuel fired boiler. This means that the nitrogen in the combustion air is no longer 'fixed' and is largely governed by the nitrogen content in the coal. In Figure 6 can be seen typical figures for a coal-fired fluidised bed which are around 250-350 Vpm. These figures are well within the limits set by the United States Environmental Protection Agency.

The third advantage is fuel flexibility. Testwork has demonstrated that the performance of a fluidised bed combustor is relatively insensitive to the type of fuel used. As described above, the bed usually contains less than 1% of carbon at any time and therefore each particle of combustible

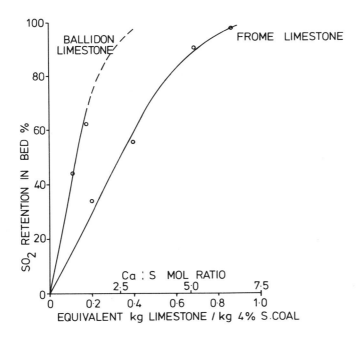

Figure 5 Effect of type and quantity of limestone on sulphur dioxide retention

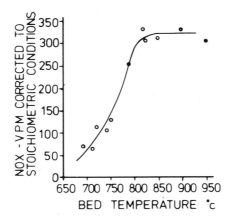

Figure 6 NO_x and bed temperature relationship firing coal with 1.1% nitrogen

material is quickly ignited by the surrounding mass of hot bed material and burnt in the hot air stream. This means that virtually any liquid, solid, or gaseous fuel can be burnt including such things as high ash coal, colliery tailings, vegetable wastes, wood chips, prepared municipal and industrial wastes, waste oils, sewage sludges, and even process offgases. Anthracite is a good example. The only limitation is that the heat released on combustion must exceed the sensible heat of the flue gases and ash produced.

The remaining features are really more of interest to the engineer than to the environmentalist. However, unless a boiler is both clean and competitive in price, it is unlikely to find a wide application. These engineering advantages include reduced fuel preparation costs compared with pulverised fuel and a more compact boiler layout because of the higher heat transfer rates which are achievable in a fluidised bed.

The other main engineering advantage is that the design of the heat exchange surface above the bed is not so dependent on the particular combustion products from the different fuels. Therefore, it should be possible to develop standard boiler designs to an extent not previously thought possible. Moreover, these designs will be flexible as to the coal used just as refineries have had to adapt to using whatever feedstock was obtainable.

As a result of the development work on fluidised combustion, which has taken place over recent years, this firing system is now starting to gain wide acceptance around the world, particularly in countries which have tight environmental standards but which are being forced to convert to coal for steam raising. For example, British fluidised bed technology is being applied by Fluidised Combustion Contractors Limited in the retrofitting of a 60,000 lb h^{-1} boiler in Ohio. This boiler shown in Figure 7 is due to be commissioned very shortly. It has been designed to use and prove the viability of using the high-sulphur coals, which are so readily available on the Eastern parts of the USA.

Fluidised bed combustion is not of course limited just to the large industrial boiler scene. There are already several small boilers in operation in the UK in the size range from 10,000 up to 30,000 lb h^{-1} and the technology is also being applied to drying of various types. At the other end of the scale design studies are currently being prepared for utility boilers rated at 200 MWe.

High-pressure Fluidised Combustion

When consideration is being given to larger power outputs the operation of the fluidised bed at elevated pressure becomes an attractive proposition. Once again the pioneering work on this process has been undertaken in Britain. The first major pilot plant of this type in the world was commissioned at the National Coal Board Coal Utilisation Research Laboratory at Leatherhead in 1969.

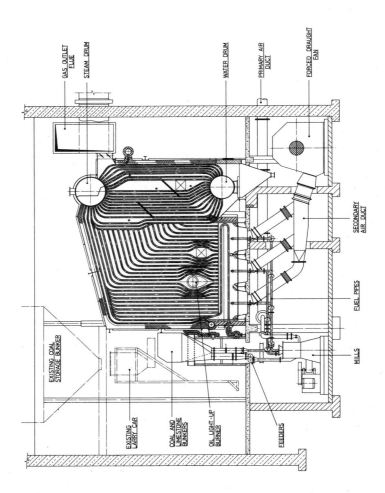

Figure 7 60,000 lb h^{-1} boiler retrofitted to fluidised bed firing

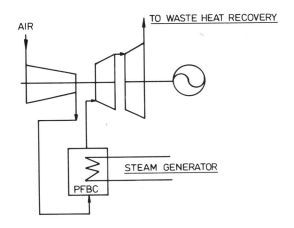

Figure 8 Proposed pressurised fluidised power generation system

As seen in Figure 8 the coal-fired fluidised bed is operated at elevated pressure, the combustion products being expanded through a gas turbine which is directly used to generate power. Heat can also be recovered from the bed using submerged tube surfaces, the high-pressure steam being used to generate additional power. This type of combustion system has all the advantages of an atmospheric fired bed and in addition offers the advantage of improved efficiency. It is estimated that cycle efficiencies of around 42% should be achieved which is significantly higher than can be achieved in conventional stations.

Following the successful operation of the pilot unit at Leatherhead, an 85 MW (thermal) prototype plant is currently being built at Grimethorpe funded by the International Energy Agency. This will be used to produce additional data on bed performance under more extreme operating conditions than are attainable at Leatherhead. Plans are also well advanced for the construction of a much larger demonstration plant, using the British technology, for American Electric Power in Ohio. This plant, which will be rated at 170 MWe will be used to confirm the suitability of the technology for utility application.

From these comments it will be realised why it is felt that fluidised bed combustion is such a major breakthrough in combustion technology. However, this is not a universal panacea for all combustion ills. There will certainly be a place for stoker and pulverised fuel fired boilers, and there will still be a place for stack gas scrubbers on very large utility boilers when these are burning high-sulphur coals. However, undoubtedly fluidised combustion has an important part to play in cleaning up our environment.

Handling of Coal and Disposal of Ash - Coal/Oil and Coal/Water Mixtures

Turning from the combustion process and the waste gases produced, there has been considerable investment over recent years in the development of improved techniques for the reception, handling, and preparation of coal for combustion. Any major dust emission in this area can be regarded as being the result of poor engineering at the outset or inadequate maintenance once the plant has been put to work. However, in certain circumstances, there may be an attraction in complete elimination of the dust problem by the use of coal/oil mixtures or coal/water slurries.

Coal/oil mixtures are a relatively new development. A coal/oil mixture is a stable suspension which is produced by the fine grinding of coal in oil, usually in the presence of a wetting agent and stabiliser. Development work in the UK and USA has shown that coal/oil mixtures can be burnt in boilers which were originally designed for oil firing. Whilst the retrofit capability was the original incentive for the development of coal/oil mixtures the delivery of coal to one's door in pumpable dust-free form is proving an added attraction. However, there is need for far more information on the operating and maintenance costs for coal/oil mixtures before they gain wide acceptance.

Coal/water slurries have been burnt successfully for many years at the power stations at Methyl and Barony. The main incentive in this case was the cost-effective use of a material which would otherwise have been dumped. However, it does have the advantage of dust suppression.

Increased use of coal with the inherent movement and handling and storage will, of course, enhance potential dust nuisance and in this context it should be emphasised that it is necessary to be alert to the extent to which this problem and others related to handling is being solved elsewhere in the world by the use of totally enclosed hydraulic systems. There is one very well known cost-effective coal pipeline in the USA. What does not receive so much attention is the extent to which these techniques are being utilised together with pneumatic methods also for hoisting coal from the pit bottom in the USSR, People's Republic of China, and here in the UK. To visit some of the major coal mines in Russia where all coal winning, underground haulage, and hoisting is hydraulic and which are therefore dust free, is an insight into how the public image of coal can be improved from the idea of something dusty and unpleasant. Crude oil at the well is not much better unless it is inside a pipe.

At the other end of the plant there is the problem of disposal of the ash, an area which is of concern to the engineer and environmentalist alike.

With power plant sites becoming more restricted and with environmental restrictions the trend is towards handling power plant ash in its most saleable form. From the various collecting points on the steam producing unit the ash is transported by a variety of means, i.e. pneumatic, hydraulic, or mechanical to storage areas where it awaits removal off site. Many factors determine which method of handling is employed for removal off site: Bottom Ash (clinker forming approx. 20% of total ash) is generally removed by road vehicles while Fly Ash (dust forming approx. 80% of total ash) is removed in a variety of ways:

(1) Dry via pipeline direct to a local processing plant
(2) Dry in road or rail tankers for processing or disposal in land reclamation schemes
(3) In a conditioned state or as a slurry to landfill schemes

The ash from a fluidised bed can also be handled either hydraulically or pneumatically depending on the ultimate method of disposal. If limestone has been added to the bed, in order to control sulphur emission, the ash will be heavily contaminated with calcium sulphate and unused calcium oxide. The potential problems of fluidised bed ash disposal have recently been assessed in order to determine if it presents a major environmental problem. The findings have in fact been very encouraging and have shown that fluidised bed ash could be used as a concrete additive, in highway basecourse stabilisation, and as a landfill material. In addition the lime content could prove of benefit in agriculture for the rectification of acidic or other deficient soils. Therefore the introduction of fluidised bed technology in no way appears to be adding to the problems of ash disposal.

It is hoped that this review has helped to give a better view of the very significant advances which have been made in the development of improved coal handling and combustion equipment. When these have been universally applied coal should no longer be regarded as a dirty fuel in comparison with oil or natural gas.

Environmental Effects of Producing Substitute Natural Gas

F. E. Dean and B. Goalby
BRITISH GAS CORPORATION, LONDON

Introduction:
Methodology and Criteria for Environmental Assessment

The authors must point out that no production coal-based SNG plant has yet been constructed in the United Kingdom, nor is one expected to be in operation before the mid-1990s at the earliest (Government Green Paper on Energy Policy). The paper must, therefore, be regarded as of a somewhat speculative nature even though the whole subject of coal gasification is receiving much attention by the British Gas Corporation and plans are in various stages of formulation. In principle efforts are being made to work out, on a contingency basis, a satisfactory distribution of numbers and sizes of SNG plants to meet the expected gas requirements in the future and to define the most suitable locations for these plants. An underlying assumption for this exercise, which may well prove to be unfounded, is that a progressive decline in availability of natural gas will call for some SNG plants to be in operation about the turn of the century.

It is not altogether satisfactory to attempt a description of the environmental effects of producing substitute natural gas without first indicating the methodology and criteria appropriate to the assessment of these effects and commenting on the consequent determination of site suitability for coal gasification plants.

The methodology is best described as dynamic in that it endeavours to carry out assessments against a moving scenario of the following activities, rather than being a once and for all static project evaluation:

(1) Site survey and measurement
(2) Site preparation
(3) Plant construction
(4) Plant operation

In principle there are five main categories of non-visual considerations involved in the methodology, together with a visual impact assessment. These five categories are geographical, technical, economic, social, and ecological. It is inherent in the nature of these parameters that it is often not possible to deal with them in an absolute sense and only infrequently can quantitative standards with tolerances be applied. In the majority of analyses the criteria have to be applied relatively and a considerable amount of subjective judgement is involved. Nevertheless, experience of the British Gas Corporation over the past 10 years, during which the methodology and criteria have been consistently applied, has shown that a high degree of correlation between the results of different practitioners can be attained.

A few comments are appropriate here on the five non-visual considerations mentioned above. The geographical aspects have to be examined first because analysis of environmental impacts cannot proceed further for specific sites for coal gasification plants if constraints on development imposed by geographical factors, natural or man-made, prove insurmountable. The technical and economic factors include not only engineering and scientific matters, but also financial considerations, of which there are many to consider where the gas industry is concerned. Pollution of land, water or air may be possibilities and an increase in ambient noise levels is also an important topic to which the general public, as well as site operators, are nowadays very sensitive. It will be appreciated that very large financial penalties may arise if it is necessary to depart too far from technical optima, as can be illustrated by moving an SNG plant away from its best position in relationship to gas pipelines and coal supply.

Several very significant social items have to be taken into account in the environmental analysis. Typical of these

Environmental Effects of Producing Substitute Natural Gas

are the effects of introducing a large workforce, often for the construction period only, on housing, social service, schools, and transport facilities. Equally important are the health and safety factors.

Ecological considerations also have to be examined where the effects of a development on the physiography, flora, and fauna of an area are involved.

It is relevant at this point to refer in more detail to the criteria as they apply at various stages of the analysis:

(1) Site Survey and Measurement

The main items in this section are as follows:
- (i) Relation of size and shape of site to size and shape of SNG plant.
- (ii) Location of site in relation to location of coal source.
- (iii) Access to site in relation to coal transport.
- (iv) Nature of site terrain.
- (v) Nature of soil in relation to building foundations.
- (vi) Site location relative to National or Regional Gas Transmission Systems.
- (vii) Relation of site under examination to projected location of other SNG plants.
- (viii) Consideration of necessity of extra compressor locations, depending on output pressure of gas from SNG plant, distance gas has to travel, diameter of pipeline, and pressure required at offtake point.
- (ix) Security of site at all stages of development.
- (x) Health and Safety aspects of site personnel and neighbourhood at all stages.

(2) Site Preparation

A certain degree of overlap between this and the preceding section occurs. Additional criteria specific to this section are:
- (i) Access of contractors' vehicles, materials and equipment.
- (ii) Availability of water, electricity, and site preparation gases.
- (iii) Temporary contractors' housing and facilities.
- (iv) Disposal and control of wastes from earthmoving and site preparation. Solid, liquid and gaseous wastes

may occur and the satisfactory control of noise and vibrations will be necessary.

The same topic occurs under Plant Construction and Plant Operation, but all these sections differ in the nature and amounts of waste and noise to be controlled. Consideration has to be given to the on- or off-site disposal of solid and liquid wastes in accordance with the Control of Pollution Act, 1974.

(3) Plant Construction

Environmental impact analysis will involve a study of the following topics:
- (i) Access of constructors' vehicles, materials, and equipment and the plant items.
- (ii) Disposal of wastes.
- (iii) Availability of utilities, e.g. water and electricity.
- (iv) Facilities for contractors.
- (v) Local infrastructure.

(4) Plant Operation

Applicable criteria are:
- (i) Methods of transport of coal to site and within the gasification plant.
- (ii) Access to site for coal, depending on method of transport.
- (iii) Coal storage.
- (iv) Techniques of coal handling.
- (v) Coal preparation.
- (vi) Utilities requirements -

  Water - Process water, boiler water, cooling water.
        - Mains, tanks, holding ponds.
  Steam raising.
  Electricity.
  Gases.

- (vii) Auxiliaries - Utility boiler
  - Wastewater treatment plant
  - Make-up water treatment plant
  - Oxygen plant
  - Sulphur removal plant
  - Standby electricity generating plant
  - Maintenance Department

(viii) Disposal and Control of Wastes:

Solid or semi-solid	Refuse from coal cleaning / Slags from gasifiers / Sludges from water treatment
Liquid effluents	The greater part, containing gas liquors, phenols, and hydrocarbons, will be trapped and gasified or sold, but a certain quantity will require a treatment plant, for which the present technology is adequate
Atmospheric emissions	Odours / Carbon dioxide from acid-gas removal plant / Sulphur dioxide from sulphur removal plant / Nitrogen from oxygen plant / Emissions from coal driers / Venting and flaring of off-specification gas
Noise	

(ix) Disposal of Saleable By-products

- Liquid hydrocarbons
- Ammonia
- Sulphur
- Phenols

(x) Storage of SNG on site

(xi) Employee Housing

- Schools
- Shops
- Medical facilities

The UK Gas Distribution and Transmission System

The British Gas system consists of 220,000 km of mains (including transmission mains) and over 14 million services.

The extent of the National High Pressure Transmission System is shown in Figure 1. Five natural gas terminals are included, about 5000 km of high-pressure transmission mains and 13 compressor stations to boost gas through the system. Regional high-pressure mains amount to about twice the length of the national grid.

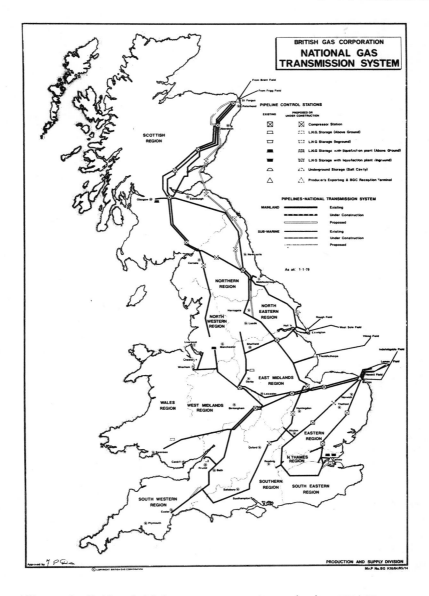

Figure 1 National high pressure gas transmission system

Coal Gasification Processes

On present evidence British Gas considers that the future direct production of SNG from coal will initially rely on the British Gas-Lurgi Slagging Gasifier which has been developed to commercial status and selected for a large-scale test programme in the United States.

As indicated in the Introduction it is not possible to say exactly when full-scale production coal gasifiers will be required, since this depends at least partly on the rate at which proven natural gas reserves will decline. The nearest one can say at present is that some SNG production from coal gasification above ground may be required to operate in the mid-1990s.

The optimum mix of plant capacities and size required to supply the entire national demand for gas with SNG cannot be accurately described today, because this depends on factors which still have to be assessed, such as the economics of plant size (scale factor), proximity to gas pipeline routes, availability and accessibility of coal, and so on. The most that can be said with confidence is that the minimum area of land required for an SNG plant making 250 million cubic feet of gas per day from coal, which is likely to be the maximum, will be about 200 acres and that for a plant making 100 million cubic feet per day about 75 acres. To produce the current consumption of gas entirely from coal gasification would call for the equivalent of 20 such plants, each of the larger size. As the British Gas Corporation does not own enough suitable operational sites to meet this requirement it will at some point in the future have to obtain additional land.

Essentially the Slagging Gasifier referred to above will take coal into the top of a cylindrical vessel, into the base of which are fed steam and oxygen. The coal is heated in a stable continuous process. Gas is evolved and incombustible residues from the coal are removed from the bottom of the fuel bed as a molten slag. Figure 2 shows the general nature of the plant.

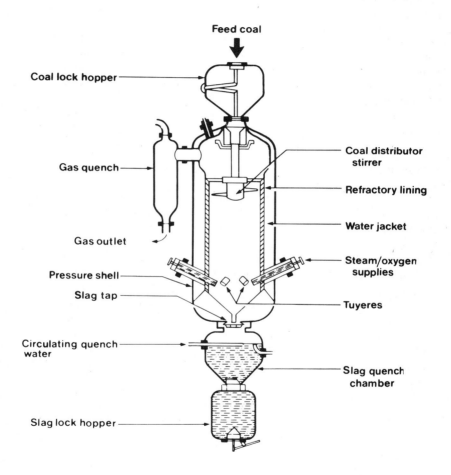

Figure 2   The British Gas/Lurgi slagging gasifier

Developments in mechanical mining by the National Coal Board are expected to produce a larger proportion of fine coal than can be readily accepted by the slagging gasifier. Accordingly the British Gas Corporation is now developing an important modification, called the Composite Gasifier, which is confidently expected to alleviate the problem of gasifying fine coal and which will go on pilot-scale production tests in the course of the next two or three years.

A typical process for producing SNG from coal is shown in Figure 3. The gas at first produced from coal, steam, and oxygen contains only a small amount of methane, the principal constituent of natural gas, and is mainly a mixture of hydrogen and oxides of carbon. It is therefore necessary for further chemical reactions to be induced which will convert these substances into methane. This is the reason for 'Methane Synthesis' or 'Methanation' being included in the diagram. Typically, hot gas from the gasifier is quenched by

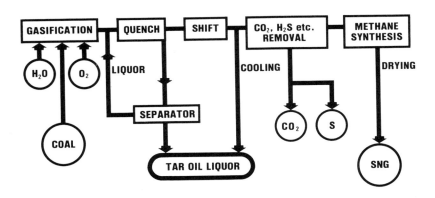

Figure 3 Typical process for producing SNG

circulating aqueous liquor. Tars are condensed, solid
particles are precipitated, and phenols, ammonia, and other
solubles dissolve in the water. The part of the diagram
marked 'Shift' is that which increases the hydrogen content
of the gas, at the expense of the carbon monoxide present,
and makes the gas suitable for presentation to the methanation
stage. Acid gases such as hydrogen sulphide and carbon
dioxide have to be removed, the former to avoid poisoning the
methanation catalysts and the latter to ensure correct
combustion properties of the final SNG.

Coal Gasification Plant

It is not particularly illuminating to show an artist's
impression of an SNG plant because, apart from the coal
handling equipment, there is a close resemblance to a petro-
chemical installation, the general nature of which is well
known. It should, however, be emphasised that an SNG plant
will be quite distinct from a plant making Town-gas from coal,
in terms of both equipment and layout. The latter will
follow the diagram which illustrates a typical gasification
process.

A key diagram of an SNG installation is shown in Figure
4 and it is by reference to diagrams such as this that one
assesses the environmental impact of the whole complex.

Pattern of Environmental Impact

It is perhaps appropriate at this point to review the general
nature of the environmental effects of producing SNG from
coal, before proceeding to a more detailed examination in
terms of the specific criteria listed in the Introduction.

Reference to the key diagram Figure 4 shows that an SNG
plant will contain many large and bulky items, some of which
have to be moved as indivisible loads. The largest size of
plant currently envisaged, namely one making 250 million
cubic feet of gas per day, would employ about 400 people full-
time. Construction of it would take 5 or more years and a
peak labour force of about 3/400 men may be anticipated.
Considerable quantities of water will be required to raise

Environmental Effects of Producing Substitute Natural Gas 137

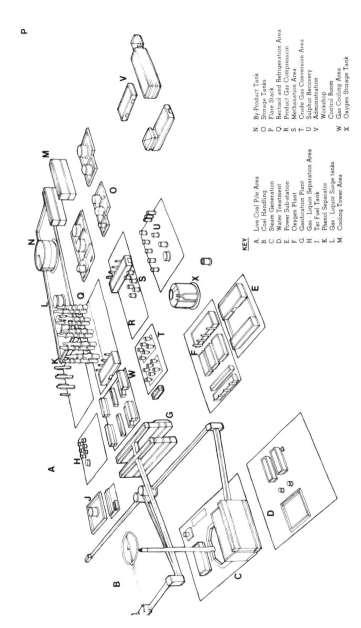

Figure 4 SNG production plant

the feedstock steam for the gasification process, there will
be a significant power requirement, and large quantities of
wastes and by-products will have to be disposed of.

It should be made clear that the only major environmental impact apart from mining and coal production which might occur will be at the point of gasification. In principle a SNG plant can be located at three different types of site:
(a) At or near a pithead
(b) At a point near to the gas transmission system
(c) In or near centres of demand for gas, perhaps using existing gasworks sites.

Environmental analysis of these three choises is rather outside the immediate purpose of this paper. It will be more useful for the purposes of this conference to attempt some assessment of the environmental effects of producing SNG in terms of the criteria sections in the Introduction and of the environmental virtues of SNG as a fuel energy source.

Application of Environmental Criteria

This paper is primarily concerned with the environmental effects of producing SNG in a fully operational plant, but it would be wrong to exclude from consideration the preparation and construction phases.

Taking the criteria listed under Section 2 of the Introduction we can note the following as being amongst the effects on the environment of site preparation:

Quite possibly new roads and access points may have to be made to enable contractors to bring on to site their own equipment. The presence of large numbers of heavy vehicles bringing materials to or from the site could be a major impact. New supplies of utilities may be necessary and it should not be forgotten that temporary buildings and car parks will have to be constructed. The Control of Pollution Act 1974 has to be considered in relation to its four parts governing solid wastes, liquid effluents, atmospheric emissions, and noise. Health and Safety of personnel on or near the site will be paramount considerations.

Plant construction, as categorised above, will involve similar criteria and pre-planning may have to include provision for further arrangements to facilitate the movement of large and heavy indivisible items of plant.

Operation of the plant is the circumstance for which the environmental effects can be most clearly defined, although they may not be capable of quantification without the application of matrix environmental impact analysis techniques. Transport of coal to the site and within the relevant parts of the gasification plant will create environmental impacts dependent on the methods used, the distances over which the feedstock is moved, and its exposure to the air. If an SNG plant is located at or near a pithead it is most likely that the coal will be moved by conveyor, so that suitable precautions to minimise noise and dust will be necessary. Rail or sea transport, on the other hand, are the most likely transport methods which would be used if the SNG plant is located near to the gas transmission system. Road transport is not favoured in this case because of the large quantities of coal to be moved. Based on present day knowledge of gasification technology the quantity of coal required to produce 250 million cubic feet per day of SNG is about 15,000 tonnes per day and this could rise to 25,000 tonnes per day for the same output of gas if lignite containing about 30% moisture were to be used. Location of an SNG plant on an existing gasworks site is most likely to involve rail transport since the movement of 15,000 tonnes of coal per day by road would be impracticable on sites in or near industrial conurbations.

Use of conveyor belts for in-plant transport inevitably produces some dust when the coal drops from one belt to another or from a belt to storage, but this can be minimised with enclosed belts and dust extraction systems.

Coal storage will probably have to accommodate a month's stock, piled to about 20 feet and compacted to guard against spontaneous combustion. A month's stock for a 250 million cubic feet per day SNG plant would occupy about 50 acres.

Run-off resulting from one inch of rain over an area of 50 acres would be about 1 million gallons. Such run-off may be highly acid and contain much dissolved matter under certain circumstances so that the treatment of it would be a matter for serious consideration.

Noise and dust associated with coal handling and preparation equipment can nowadays be held at a low level by correct design and screening and none of the problems should differ greatly from those successfully tackled by the CEGB.

The provision of the utilities for the gasification plant is, like the provision of the coal, of the highest importance. Water is required in large quantities to generate steam for use as a feedstock quite apart from other normal uses and it is estimated that a 250 million cubic feet of gas per day plant will require about 8 million gallons of water per day. Nowadays it is possible to purify fairly low-grade waters sufficiently well for boiler feed by reverse osmosis. Water recovery systems are used where possible in an SNG plant to reduce the amount required so that overall less water is required than for a power station of equivalent energy output. A large quantity of water is produced as a by-product of the methanation reaction, approximately 3000 gallons per million cubic feet of gas. This water is largely uncontaminated, except for dissolved carbon dioxide which is easily removed. After suitable treatment the water can be re-used for the boiler feed.

The provision of adequate electricity and gas supplies is also necessary, together with the auxiliaries listed in the criteria in the Introduction. The environmental effects of these will require assessment, particularly of the larger items of plant, such as for oxygen production, sulphur removal and treatment of waste waters.

Proper environmental control of waste residues, including those from coal gasification and combustion, is, of course, the subject of extensive legislation of which Part I of the Control of Pollution Act is possibly the most important. This is concerned with solid (or semi-solid) wastes. For a typical SNG plant the main solid waste materials to be removed from site per day will be about 2000 tonnes of slag, derived from the coal ash, about 200-300 tonnes of elemental sulphur and about 10 tonnes of sludge from water and effluent treatment. An additional 100 tonnes of solids may also arise if coal is used as fuel for steam raising. Some of this slag could be marketed for road-fill or other purposes but what cannot be marketed should present no environmental problems since it is eminently suitable for general landfill purposes.

It should be possible to find markets for the sulphur as it will be of high quality. The sludge from effluent treatment is likely to contain most of the trace elements since these would tend to concentrate in it. This may have to be disposed of as waste, although some may have a significant value in the future. If coal were used for steam raising the ash and particulate matter would require disposal and again most would go for landfill purposes.

As previously mentioned the greater part of the liquid effluents can be collected and processed to produce valuable by-products based on such compounds as ammonia, phenols, and hydrocarbons. The British Gas Corporation has developed satisfactory treatments for those constituents of the effluents which cannot be handled in this way.

The main potential sources of particulate and gaseous emissions are steam raising plant, acid gas removal units and liquefaction plant. The combustion products from coal if used for steam raising would be a source of gaseous pollution. Thus a 250 mcfd SNG plant could generate about 1½ tonnes per hour of sulphur dioxide, which quantity is usually regarded as dispersable by the use of high stacks. It would, however, be removable by one of the many packages now available for flue gas desulphurisation but the economics of these processes are dubious. Removal of particular matter from coal combustion gases should present no major problem because the existing and well proven cyclone and electrostatic precipitation technology is sufficient to meet present and likely pollution legislation.

One other main potential source of gaseous pollutants is dealt with at the acid gas removal stage. The gasification of coal containing 1.6% sulphur will produce about 8 tonnes per hour of sulphur-containing gases, but the removal of these together with the carbon dioxide is not expected to create a major problem since the required techniques are already available.

Where noise is concerned there are likely to be many sources including coal handling equipment, compressors, vents, and high-pressure gas lines. However, the coal equipment would not be any noisier than that used in power stations and would be silenced to a similar level. For the other items British Gas Corporation has very considerable experience in controlling noise from sites such as compressor stations and in pipelines installations to readily acceptable levels.

It would be inappropriate to close this section of the paper without emphasising the importance of employee housing, schools, shops, and medical facilities, which constitute the local infrastructure vital for projects of this type.

Environmental Virtues of SNG

Production of SNG represents the most efficient way of extracting energy from coal other than by burning it. Listed below are strong environmental reasons for making this type of gas:
  (i) The gas is non-toxic.
 (ii) Its combustion is essentially a non-polluting process.
(iii) It can be transported very efficiently by existing high-pressure pipelines with negligible effect on local inhabitants.
 (iv) Storage in gaseous or liquid form enables large amounts of energy to be safely and economically handled.
  (v) As the only major impact, if at all, is at the point of gasification, the overall net environmental effect of SNG production, in the sense of 'coal to usable heat' transformation, is considerably less than by any other coal-based route.

Environmental Effects of Producing Substitute Natural Gas

The authors wish to thank the British Gas Corporation for permission to publish this paper, but emphasise that the views expressed are theirs alone and do not necessarily present British Gas policy.

...iction of Liquid Fuels from Coal

Gibson
NATIONAL COAL BOARD, LONDON
D. W. Gill
COAL RESEARCH ESTABLISHMENT, NCB, STOKE ORCHARD, GLOUCESTERSHIRE

Introduction

Since the middle of the 18th Century coal has been processed on a commercial scale for the manufacture of secondary fuels, required for special uses, such as coke for blast furnace and foundry operation, coal gas for lighting, heating, and cooking, and smokeless fuels for domestic heating. The by-products of processing have been used in chemicals, fertiliser and explosives manufacture, and more recently in the building and construction industries.

Before the end of the century, the world demand for oil is expected to outstrip production capacity, causing severe price rises, and this will lead to a situation in which liquid fuels derived from coal can supplement natural oil supplies on a comparable economic basis. Reserves of coal greatly exceed those of oil, and in due course coal is likely to become the major source of liquid fuels required mainly for use in transport. By-products from petroleum refining are used at present as a source of organic chemicals and plastics, and as oil reserves dwindle these materials will increasingly be made from coal.

There are two routes by which coal can be converted into liquid fuels. Both routes result in an increase in the ratio of hydrogen to carbon, and elimination of the mineral matter (ash) associated with the coal, together with much of the sulphur and other minor and trace elements in the coal. The first route, known as the synthesis route, involves gasification of the coal to produce raw synthesis gas (mainly carbon monoxide and hydrogen), purification of this gas, adjustment of the composition to suit the products required, and reaction in the presence of catalysts to synthesise the organic fuel molecules. The Fischer-Tropsch is the best known of the processes using this route, and Table 1 shows the reactions involved. The second route involves degradation of the coal substance, either by heat alone or in the presence of an organic solvent. The hydrogen is added during or after degradation, either as hydrogen gas or as hydrogen donor substance which can give up part of its combined hydrogen to the coal products. Many processes in this category also employ a catalyst to facilitate hydrogenation, and the crude oils produced are then subjected to a further catalytic hydrocracking similar to that used in the refining of crude natural oil.

Table 1   Fischer-Tropsch reactions and mechanism

$$nCO + 2nH_2 \rightarrow (-CH_2-)_n + nH_2O \quad \ldots\ldots +39.4 \text{ kcal mol}^{-1}$$
$$2nCO + nH_2 \rightarrow (-CH_2-)_n + nCO_2 \quad \ldots\ldots +48.9 \text{ kcal mol}^{-1}$$
$$nCO + 2nH_2 \rightarrow H(-CH_2-)_nOH + (n-1)H_2O \ldots\ldots +58.5 \text{ kcal mol}^{-1}$$

Water-gas shift $CO + H_2O \rightleftharpoons H_2 + CO_2 \ldots\ldots +9.5 \text{ kcal mol}^{-1}$

$$\underset{M(CO)_x}{R\backslash} \xrightarrow{+CO} \underset{M(CO)_x}{R-CO\backslash} \xrightarrow[-H_2O]{+2H_2} \underset{M(CO)_x}{R-CH_2\backslash} \xrightarrow{+CO} \underset{M(CO)_x}{R-CH_2-CO\backslash}$$

The first production of liquid fuels from coal was by the distillation of coal tar from coke oven or gas works operation. Yields were small, and only represented a minor by-product of the main yield of coke or gas (Table 2). The first plants for the commercial manufacture of liquid fuels as the main product from coal were built during the early 1930s in many European countries, including Britain, where ICI operated a 100,000 tonnes per annum coal hydrogenation plant at Billingham for four years before it was closed on economic grounds. Many plants were built in Germany during the Second World War to compensate for the loss of oil imports resulting from blockading of sea routes. Twelve coal hydrogenation plants were built, based on the Bergius process, and these produced more than 3 million tonnes of oil per annum during 1943-44. A

Table 2 Products of high-temperature carbonisation
        (900-1100 °C)

Product	% Wt.
Gas	18
Light oil	1.0
Tar	4.5
Coke	75

further half a million tonnes per annum of liquid fuels and natural fats substitutes were also being produced in Fischer-Tropsch synthesis plants, but towards the end of the war most of the plants were destroyed by Allied bombing raids.

Since the war, most of the development in coal hydrogenation processes has been in the USA, where three processes are nearing the stage of commercial application. These are the H-Coal, Exxon Donor Solvent, and Solvent Refined Coal processes. Several other processes are also being developed or are in abeyance, both in the USA and in Germany. In addition, new processes based on pyrolysis and on synthesis are being developed, such as the COED pyrolysis process and the Mobil synthesis process. Of the older processes, only the Fischer-Tropsch is commercially operated today, principally in South Africa.

In Britain, work is being conducted by the NCB at the Coal Research Establishment at Stoke Orchard. Two variants of the degradation/hydrogenation route are being investigated. One process involves liquid extraction of the coal using a process-derived solvent which behaves as a hydrogen donor. A flow diagram of this process is shown in Figure 1. The other process uses extraction by a gaseous hydrocarbon at supercritical temperature and pressure. This allows easier and more efficient separation from the undissolved part of the coal. A flow diagram is shown in Figure 2. Both processes produce a solid, carbonaceous residue, which also contains almost all of the mineral content of the coal. The supercritical gas extraction residue is highly reactive and can be gasified to produce hydrogen and fuel gas for use in the process, or it can be used as a fuel for steam-raising.

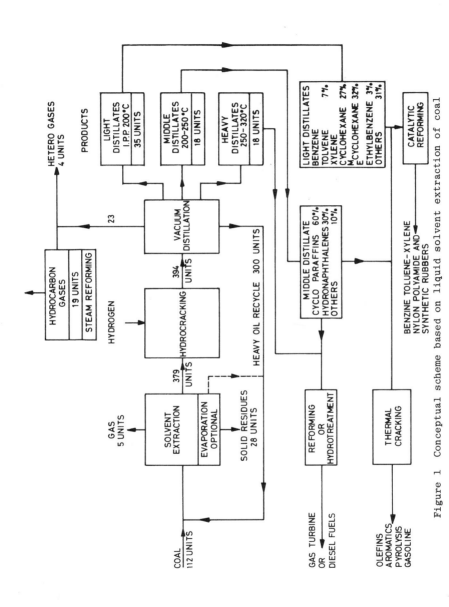

Figure 1 Conceptual scheme based on liquid solvent extraction of coal

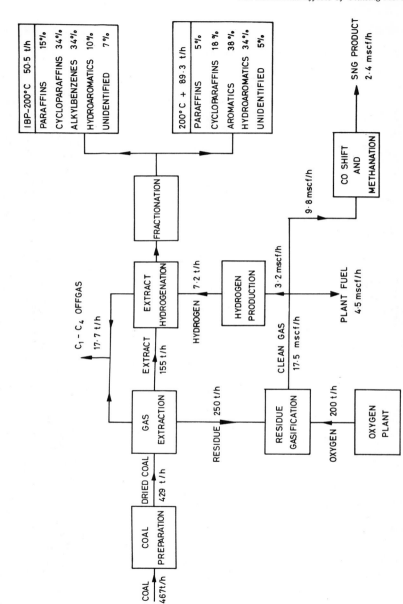

Figure 2 Conceptual scheme based on gas extraction of coal and SNG production from the residue

Environmental Impact

We shall describe a conceptual full-scale production plant (10,000 tonnes a day coal input) based on a simplified form of two-stage liquid extraction process in which the coal is first extracted in a solvent, the extract then being hydrogenated. There are many possible process variations, and it must not be inferred that this concept represents the optimum detailed arrangement for a production plant. Experience on the pilot plant scale may show that a different variant is preferable, although it will probably still be a two-stage process. We will look especially at the demands made on resources in building and operating the plant, and at the impact of the plant on the environment. There will be some loss of amenity to residents living near to the plant due to noise, visual impact of the plant, and nuisance from road and rail transport entering and leaving the site. There will also be concern about possible harm to the environment due to the emission of gaseous and liquid effluents, and we shall discuss the forms of treatment that can be used to remove harmful constituents of the effluents.

Siting of the plant is the first consideration, and the choice will probably rest between a site adjacent to a colliery expected to remain in production for a long time, and one close to an existing oil refinery, where expertise and facilities will be available for the cracking and refining steps to produce a range of saleable products. If the choice is the latter, coal may have to be brought by rail some distance from the point of production, on a scale comparable to that required by a large power station. A colliery site would be preferable environmentally, because coal delivery could be by conveyor belt and the oil products could be sent by pipeline to an existing distribution depot.

The area of land needed is about 50 hectares, or 14 hectares per million tonnes per annum coal input. This is appreciably more than needed for an oil refinery, but well under that for a power station of the same coal input. An artist's impression of the plant alongside a 2000 MW power station on the same scale is shown in Figure 3. Some

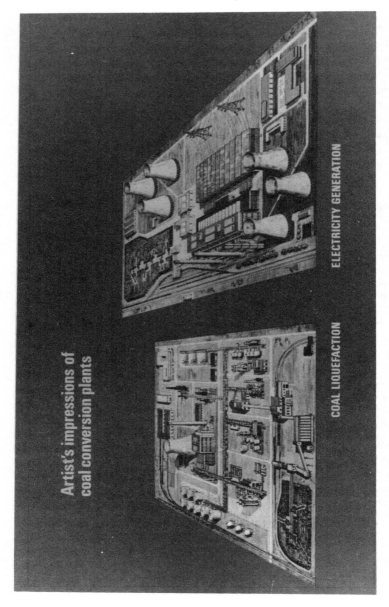

Figure 3 Comparison of artist's impressions of a coal liquefaction plant and a power station. The visual impact of a coal liquefaction plant should be no greater than that of a power station of equivalent energy output rating, in these cases 2000 MW

indication of the demands on resources for construction of the plant is given by the estimated capital costs. At between £95 and £125 per tonne per annum of coal input (mid-1979 prices) these are comparable with costs of a coal-fired power station of equal energy input, and a good deal less than those of a nuclear power station of the same rating.

A simplified diagram of the entire plant is shown in Figure 4. In this example, the coal input is split into three streams, two of which are dried and prepared to a suitable particle size. One stream passes to the solvent extraction plant and the other to a gasification plant to provide fuel gas for process heating and hydrogen for hydrogenation of the extract. The third stream passes straight to the electric power plant without drying and is supplemented by solid residues from the extraction stage. After hydrogenation, the extract is fractionated and fractions boiling above 250 °C are recycled as solvent. To avoid a build up of very high molecular weight material, a proportion of the recycle stream is diverted to a coker where it is pyrolysed to give an oil which is recycled to the digester and coke which forms a starting material for the manufacture of high-grade carbon products - furnace electrodes etc.

Impure hydrocarbon gases are evolved at various stages of the process and these are treated to remove hydrogen sulphide; then hydrogen is separated cryogenically for use in hydrogenation of extract. The rest of the gases are fractionated to give substitute natural gas and LPG, for which there will be markets outside the plant.

The range of chemical compositions of the crude coal extract (before hydrogenation) and of coal is shown in Figure 5. The range of sulphur contents is greatly reduced, but that of nitrogen not greatly altered. Hydrogenation results in a further reduction in sulphur to less than 0.3%. Sulphur is present in coal mainly in two forms: pyrites and other metal sulphides, and as organically combined sulphur. The sulphide sulphur remains in the solid residue from extraction and passes to the boilers where it is oxidised to sulphur dioxide.

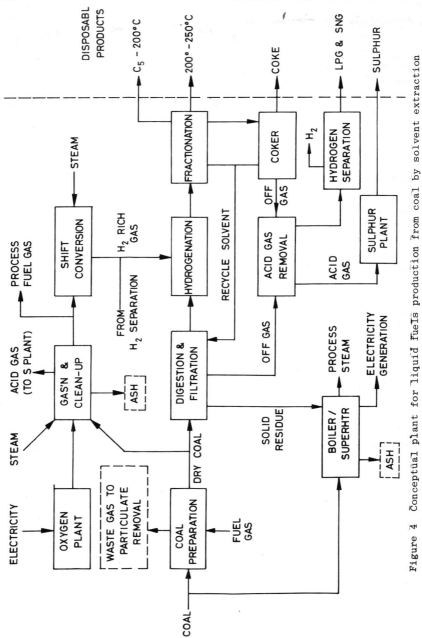

Figure 4 Conceptual plant for liquid fuels production from coal by solvent extraction

In a pulverised-fuel fired boiler this sulphur dioxide would normally be emitted from the chimneys, but commercial processes for flue gas desulphurisation are now available. If a fluidised bed fired boiler were installed, up to 90% of the sulphur in the fuel could be retained in the bed by the addition of limestone or dolomite.

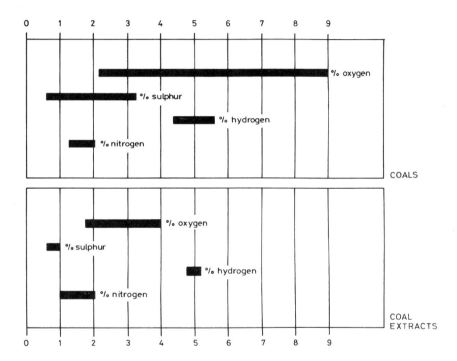

Figure 5   The composition of coals and coal extracts. The transformation of coals into coal extracts is accompanied by the elimination of some components, particularly oxygen and sulphur. Although the elemental composition of coals can be extremely varied, the range of values for the composition of coal extracts is fairly limited. Coal extracts are much less variable in composition than crude oils.

The organically combined sulphur in the coal is mostly evolved with the off-gases from digestion and coking, where it appears as hydrogen sulphide. One of several available processes for acid gas removal would be used to remove hydrogen sulphide from these off-gases, and also from the cooled gas from the gasifier and from the foul gases resulting from sour water stripping. Recovery of the sulphur would be as elemental sulphur or as sulphuric acid if there were a demand locally for this chemical.

Sulphur production from operation of the plant on a British coal of average sulphur content would be about 30,000 tonnes a year in the absence of boiler flue gas desulphurisation (FGD), or 34,000 tonnes a year if a regenerable FGD system were applied on the boiler flue gases. In the absence of any measures to remove sulphur oxides from the boiler flue gases, about 7500 tonnes of SO_2 would be emitted annually. For comparison, the emission from a 2000 MW power station on 60% load factor is over 100,000 tonnes a year. Other boiler plant emissions can be controlled by techniques currently available, or being developed as, for example, combustion modifications for NO_x reduction.

An important environmental aspect is the water demand for the plant and the disposal of waste water from the process.

Water will be required at various places in the process: make-up boiler water to replace steam used in the gasifier; shift conversion and sour-water stripping; and as an input in the gas/liquid separation stage of the extraction plant where it will dissolve ammonia and hydrogen chloride. This contaminated water will join the waste water from the gasification plant which will contain dissolved and suspended solids together with a mixture of organic compounds including phenols and tars.

The combined foul water will be treated by steam-stripping, with recovery of ammonia and phenols, sedimentation, biological purification, filtration, and, if required, by ion-exchange demineralisation and activated-carbon treatment. The demands for cooling water, largely for steam condensers in the electricity plant, will almost certainly exceed the foul water flow rate, and a very high standard of purification will not normally be called for to give water of a quality suitable for cooling. Much of the water used for cooling is lost by evaporation, and this greatly reduces the volume entering the final treatment stage.

A generous estimate of demands for make-up water on the site is 220 litres per GJ output, which is about a quarter of that for a power station of equivalent energy output. A recent American paper[1] has estimated that by careful re-use of water and minimisation of the proportion of wet cooling used it should be possible to reduce water demands by 60 to 80% compared with earlier estimates on which the figure of 220 litres per GJ was based.

Solid effluents from the plant will be mostly coal ash from the boilers and the gasifier, residues from water treatment, and small amounts of spent catalyst from time to time, when regeneration is no longer possible. The ash, amounting to between 40 and 50 tonnes an hour on the conceptual plant, will be free from organic compounds and will not present any health hazards in handling. It could find outlets in the manufacture of building products and aggregates, or as an infill in civil engineering projects. The spent catalyst will be reactive and will have to be specially processed before disposal. Many of the substances used as catalysts contain scarce and valuable metals, and in many cases return to the manufacturers for recycling will be economic.

Sources of emission of particulate matter are likely to arise from combustion of coal and extraction residues in the boiler plant, as well as from coal storage, preparation, and transport within the plant.

Combustion-derived particulates emission will be subject to a limit agreed with the Alkali Inspectorate, with whom the plant will be required to be registered. The limit will probably be in line with that applied to other large coal-fired power plant; this is currently 115 mg per cubic metre of dry gas measured at 288 K and expressed at a reference CO_2 content of 12%. It corresponds to an emission of just under a kilogram of ash for each tonne of coal burnt.

The emission of fine coal particles from the coal drying and grinding processes is one of the problems encountered at present in the operation of smokeless fuel and other coal conversion plant. The method of control generally used is a single-stage electrostatic precipitator, which collects over 95% of the dust entering it and reduces the emission to about 200 g for each tonne of coal processed.

However, in a new plant of the size contemplated for liquid fuels production it will probably be desirable to aim at a still lower emission rate from the coal dryers, and this could be done by using two-stage or three-stage electrostatic precipitators or, alternatively, afterburners.

Dust from the coal storage area may be minimised by the use of shelter-fences or tree screens to reduce wind velocities, by spraying stockpiles with emulsions of a surface binder, and by good house-keeping generally, including regular sweeping of roadways and working areas. Fugitive dust from coal transport will be reduced by careful design of conveyor systems, particularly in respect of the hooding of transfer and discharge points.

Nature of the Products

Some concern exists that the liquid fuels produced by the degradation and hydrogenation of coal may be more physiologically active than those made from natural crude oil and may therefore present a health hazard to process workers and people handling the products. Little definite information exists on this question, but it is a fact that the aromatic and cyclic hydrocarbon contents of the coal-derived fuels are higher - about 75% compared with 35% in petroleum. Measurements of the benzo(a)pyrene content of a sample of the crude product of a degradation coal liquefaction process showed only 50 p.p.m., which is much closer to that of petroleum (30 p.p.m.) than to that of coal tar (2,000 to 20,000 p.p.m.).[2] Benzo(a)pyrene is often taken as an indicator of the group of polynuclear aromatic hydrocarbons that possess carcinogenic activity.

Coal-derived liquid fuels also contain a considerably greater nitrogen content than do petroleum products, and this fact has also given rise to concern because of the known physiological action of many organic nitrogen-containing compounds. Hydrogenation results in a considerable reduction in nitrogen content, to less than half that in the crude extract. Workers at the US Oak Ridge National Laboratory have tested fractions of a crude coal liquid (Synthoil) for mutagenic action.[3] Activity was found in the fraction containing indole and related compounds, as well as in the fraction containing polynuclear aromatic hydrocarbons. However, no comparative tests on crude natural oil fractions were reported, so the significance of the results is not clear. It should be noted that the higher aromatic content of coal-derived fuels reduces the need for lead alkyl additives in motor spirit, and these additives are probably far more dangerous to health than the minor constituents of the refined oils from coal.

Energy Efficiency

Fuel conversion processes inevitably entail some loss of energy, dissipated as heat generated on site and lost to the atmosphere and outgoing water streams. Coal liquefaction is no exception, and an overall energy efficiency may be evaluated as the total potential energy in the saleable products divided by the potential energy in the coal entering the process.

Estimates for the NCB liquid solvent extraction plant show an overall energy efficiency of between 65 and 70%, roughly two thirds of the energy output being in the liquid fuels with boiling points up to 250 °C and the rest being in SNG, LPG, and coke. The expected efficiency compares favourably with those found in practice for processes operating by the synthesis route, e.g. 55% on the SASOL plant which has been operating at Sasolburg in South Africa since 1955, and less than this in the design for a second SASOL plant aimed at producing principally aviation and motor fuel.

The energy efficiency, and also the general environmental impact, of the gas extraction process under development by the NCB are expected to be similar to those of the liquid extraction process which has been dealt with in detail in this paper. Approval has been given for the design of 1 ton h^{-1} pilot plants for the further development of both processes, and Figures 6 and 7 show the proposed schemes for these pilot plants. After much consideration of 16 possible sites in Britain, Point of Ayr in North Wales has been chosen for both plants.

Acknowledgements

The authors gratefully acknowledge the help and advice given by their many colleagues within the National Coal Board in the preparation of this paper. The views expressed are those of the authors and not necessarily those of the Board.

References

[1] H. Gold, J.A. Nardella, and C.A. Vogel, 'Fuel Conversion and its Environmental Effects', CEP, August 1979, pp.58-64.

[2] L. Parker and D.I. Dykstra, eds., 'Environmental Assessment Data Base for Coal Liquefaction Technology: Vol. II', Report No. EPA-600/7-78-184b, September 1978.

[3] C.-L. Ho, B.R. Clark, M.R. Guerin, C.Y. Ma, and T.K. Rao, 'Aromatic Nitrogen Compounds in Fossil Fuels - a Potential Hazard?', ACS, Div. Fuel Chem., Preprints 1979, 24 (1), 281-291.

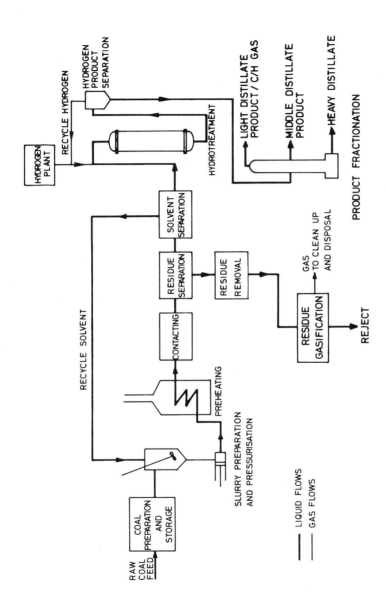

Figure 6 Pilot-scale plant for coal liquefaction supercritical gas extraction

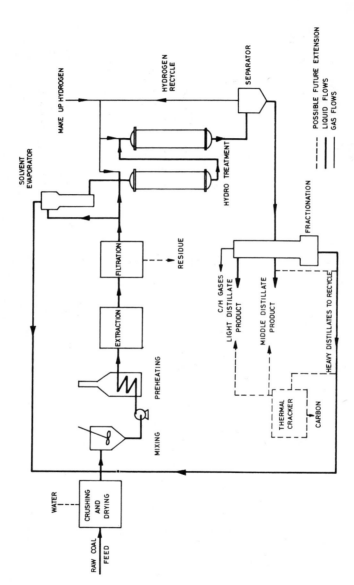

Figure 7 Pilot-scale plant for coal liquefaction using liquid solvents

Discussion on Session III and Chairman's Closing Remarks

Chairman: What efficiencies can be attained in fluidised bed combustion?
P C Finlayson: Up to 70%, the improvement being achieved by the use of very low-grade heat.

M F Tunnicliffe (Alkali & Clean Air Inspectorate): Can Mr Kindersley amplify his comment that the particulate matter in the waste gases from fluidised-bed combustion would need to be removed by using bag filters or electrical precipitators, and does he consider that this would apply also to the smaller installations?
T H Kindersley: Yes. The sulphur will be caught before the gases leave the boiler, but the particulates will still have to be removed.

G V Day (UKAEA): Does the speaker envisage fluidised-bed boilers being introduced into urban areas for use in medium and light industry and possibly commercial buildings? What height stack would be needed to meet Mr Tunnicliffe's (Alkali Inspectorate) requirements, or local government regulations? How will the problem of solid waste disposal be overcome, bearing in mind the opposition to 'waste dumping' and that the CEGB can only sell 40% of their ash?
T H Kindersley: The height of the stack will be no greater than for conventional boilers. Although it may sometimes be difficult to make profit from the sale of ash, there are many ways of using or disposing of it.
G V Day: What would be the lower end of the size range of fluidised-bed boilers likely to be installed in urban areas?

P C Finlayson: The smallest size of boiler envisaged would be capable of 10,000-30,000 lb of steam per hour.

A Keddie (Warren Spring Laboratory): Dr Goalby has referred to the 'non-polluting' characteristics of SNG at the point of use (combustion). If one said 'relatively non-polluting' then it would be easier to agree with him because natural gas does produce carbon monoxide and oxides of nitrogen during combustion. Could Dr Goalby please tell us if SNG would produce the same quantities of these gases?
B Goalby: Yes, SNG would produce the same gases as natural gases on combustion.

D G Harvey (Kennedy & Donkin): Is the amount of coal adequate for the power stations planned? It would appear that 100 m tonnes per annum would be required, and this seems a forlorn hope.
B Goalby: We have reserves of coal in the UK sufficient for some 300 years, whereas we have much smaller reserves of natural gas. The NCB is very concerned about mining sufficient coal, hence the Vale of Belvoir enquiry, but it believes its objectives will be achieved.

R E Pegg (Esso Petroleum Co Ltd): 1) What is the stability of the crude liquid products? This is important in siting the plant at the mine or at a port (refinery) location. 2) Is it better to import crude coal liquids rather than coal? Again, coal liquid stability is a factor.
J Gibson: Solvent refined coal has been transported across the United States of America from Fort Lewis on the West Coast and this was burnt successfully in power stations in the East. The liquids become more stable as more hydrogen is added. The question whether we import coal or an intermediate could become very important in a national emergency. On economic grounds alone one would not want to import coal but there are

Discussion on Session III and Chairman's Closing Remarks 163

many other factors to be considered, e.g. strategic, availability of infrastructure and refinery capacity, and so on.

G V Day (UKAEA): Substantial increases in coal consumption could be achieved by central processing of coal to produce SNG and liquid fuels, and also by the installation of local fluidised-bed boilers in industry and possibly commerce. Would the speakers care to give an assessment of the comparative national environmental impact of centralised processing in comparison with burning coal in a multiplicity of fluidised-bed boilers in urban areas? In addition, what would be the approximate extra capital and operating cost as a result of the necessary environmental controls for central coal processing plants? As an extension of this question, and bearing in mind that these plants will not be operating on any significant scale before 2000, by which time our standard of living will be hopefully much higher and quite probably our demands for environmental controls will be much more stringent. Would it be technically feasible to design these plants for zero emission, and by very roughly what proportion would the capital cost be increased?

J Gibson: We are preparing for nationwide distribution of coal by improved methods up to about 40 m tonnes of coal per annum by 2000 for industrial use with full environmental control. The capital cost for zero emissions from coal processing plants (for liquids and SNG) is built in to the total cost and should not be considered as an extra. Such plants will be high-integrity facilities operating under the strictest controls.

D W Gill: I agree that for environmental purposes it would be better to have centralised processing, as it is uneconomic to have environmental control over a lot of small equipment.

K Mellanby: In fluidised-bed combustion, it is apparently essential to add calcium carbonate to trap the SO_2. Does not this produce additional CO_2? Is underground gasification of potential importance for this country?

P C Finlayson: Some CO_2 is produced by reaction with SO_2, but the amounts are small. Most of the calcium carbonate remains as carbonates.

J Gibson: Underground gasification is not practicable in this country with present techniques. The gas produced would have a low thermal rating and to fire an average-sized power station about 10 square miles would have to be isolated for 25 years or more. A large number of boreholes would be required, and it would be difficult to fracture the coal without fracturing the adjacent strata. The position is different in the USA, where thick seams are often continuous, below open country, and underground gasification is more feasible.

B Lees: It is proposed that the NCB will carry out research programmes to produce liquid fuels from coal. Are we working in co-operation with the Americans, South Africans, and Australians to minimise duplication of effort? Are we considering hydrogenation of Australian Brown Coal as vast deposits are available and it was found some 40 years ago in experiments at the Fuel Research Station that this fuel was particularly amenable to satisfactory reaction in the Bergius Process because of the consistently low ash content? Finally, would it not be better to install the liquefaction plant on site in Australia rather than in this small country with its resultant pollution problems?

J Gibson: I agree that international co-operation to avoid duplication is desirable. Co-operation with the USA, Australia, and other countries, 19 in all, through the International Energy Agency is in fact excellent and work is going on in many countries at the same time with fairly free exchange of information. Brown coal is more difficult to hydrogenate than ordinary coal as it has a higher oxygen content. There is much to be said for having liquefaction plants in Australia rather than shipping coal over vast distances for processing say in the United Kingdom.

Discussion on Session III and Chairman's Closing Remarks

E S Rubin: Yesterday's discussion took a rather different course from this morning's and we now appear to be contemplating the removal of 90% of the sulphur.
D W Gill: The UK electricity supply industry has adopted the highly efficient 'tall stack' policy for control of atmospheric pollution from power stations, and it would be very expensive to install and operate plant for the reduction of sulphur dioxide emissions. However, in the development of new processes for the conversion or combustion of coal, such as we are discussing this morning, sulphur removal can be incorporated relatively cheaply as an integral part of the process. In the case of liquefaction, sulphur must generally be removed to give an acceptable quality of product.

S Wallin (Warren Spring Laboratory): In fluidised-bed combustion, is lime added automatically?
P C Finlayson: No, limestone or dolomite is only added to the extent necessary to capture the required quantity of sulphur. However, if the coal contains little sulphur and ash, it may be necessary to add some other material such as sand to maintain the bed height.
   It is important to remember that pollution control is only one of the features of fluidised-bed combustion. In many countries, there is more interest in its capability to burn low-grade ash coals cheaply and efficiently.

J Gibson: British coal contains only 1.5% of sulphur on average whereas American coal contains more than 3.5%; hence the stricter legislation in the USA. Therefore it seems silly to apply the same legislation in the UK with our low-sulphur coals.
F E Dean: We should surely be designing equipment to meet British and not US standards.
T H Kindersley: But we must, however, retain flexibility in designing of equipment because of the importance of the export possibilities and selling know-how.

K C C Bancroft (Plymouth Polytechnic): Yesterday we saw some electronmicrographs of powdered coal and fly ash produced from power stations. The fly ash was seen to have a glossy surface and was said to be quite safe as far as diseases such as silicosis are concerned. We have heard this morning that the fluidised-bed process operates at a lower temperature than the pulverised fuel fired power stations. It seems likely that the ash produced will not be glossy. Is there therefore likely to be a significantly increased health risk from particulate matter from fluidised-bed plant?

D W Gill: This is an important point which should be investigated. Ash from fluidised-bed combustion is much coarser than that from pulverised fuel and therefore more easily separated from the combustion gases before discharge to atmosphere.

Chairman:

We have just heard three speakers in the session giving fascinating accounts of the technology, the plans, and the costs of utilising that major fossil fuel, coal, which we in the UK are fortunate to have in such abundance: in such abundance not only on land, but apparently under the sea bed as well. (I understand that embarrassingly large deposits of coal have been encountered during the process of drilling for oil in the North Sea.)

It is vastly encouraging therefore to have heard of new methods of extracting the heat energy from coal by improved combustion techniques, and of the carefully laid plans to develop step by step from the laboratory scale to full-scale commercial plants, safe and acceptable means of extracting natural gas and liquid fuel from coal as a fuel stock. It came as a surprise to me, and perhaps to some of you, to hear of the sheer size of a single plant as envisaged for the 1990s to yield oil from coal. Such a plant raises a multitude of environmental issues of visual, aural, and nasal obtrusions, if I can put it that way, handling of fuel stock, disposal of waste products, and transport of the principal outputs of the plant.

It is good to know that all these aspects are being considered in good time and that these plants when they exist in number in the early part of the next century should be wholly acceptable to the citizens of the country at that time.

May I therefore thank our speakers and those who took part in the discussion for giving us such a valuable morning.

Session IV

Chairman's Opening Remarks

F. A. Robinson
CHAIRMAN, COUNCIL FOR ENVIRONMENTAL SCIENCE AND ENGINEERING

I am sure that Lord Ashby will regret not being able to be with us. He is very interested in this subject as you all know and those who have had experience of him realise that he is a very stimulating chairman who can drag things out of people even if they are reluctant to say anything. I would like to welcome Mr Davison here this afternoon to talk on the subject of "Some environmental aspects of solid smokeless fuel". Mr Davison says he is the only non-engineer, non-scientist to address this meeting. He is in fact an economist so we have a new viewpoint which I imagine will come across in the discussion. He has himself written a book on the environment so environmental problems are not by any means foreign to his experience. He graduated at the London School of Economics in economics and is a member of the Institute of Personnel Management and a Fellow of the Royal Society of Arts. He joined the NCB as an Administrative Assistant in 1956 and after moving to London he held posts in the Board's Industrial Relations Department and Central Secretariat. He became a Staff Officer to Lord Robens in 1964 and in 1967 joined the Board's Opencast Executive. He joined NCB (Coal Products) Limited in June 1976 when he also became Managing Director of NCB Hydrocarbons Ltd and Deputy Chairman of National Smokeless Fuels Ltd, so he has a lot of qualifications for giving this talk and then he's also Chairman of NCB (IEA Grimethorpe) Ltd, which is responsible for the International Energy Agency fluidised-bed combustion project and one or two other things. So without going any further into his history, I'll ask Mr Davison to give his talk.

Some Environmental Aspects of the Production of Solid Smokeless Fuel

D. J. Davison

NCB (COAL PRODUCTS) LTD., HARROW, MIDDLESEX

Mr Chairman, Ladies and Gentlemen, I am grateful for your not going further into my history - one is never quite sure what might emerge! When I received the invitation to come and speak to this conference I must admit I was somewhat surprised. As you heard, I am neither a scientist nor an engineer and as far as I could see this was a body representing 21, if I have got the number right, other bodies of scientists and engineers so I thought 'Well, why on earth has he invited me,' and I must admit the thought crossed my mind that your motive might have been to provide a living illustration of the type of scientific illiteracy that you have to put up with in your day to day activities and that you wanted a convenient Aunt Sally on the final afternoon of your conference so that you could let your hair down and really throw things about. If so, I would be delighted, so that I can throw one or two things back. It does seem to me, having had to deal with scientists and engineers in a quasi-professional capacity for a number of years now, that they often live in peculiar types of glass houses and I am not averse to hurling one or two stones through them. Let me make a comment about why I should come and have things thrown at me. I came because it struck me that your conference was of the greatest importance!

We have gone through periods in this country over the last 25 or 30 years in which there have been many arguments about what were the coming fuels. Do you remember - we are now in 1979 - do you remember the oil campaign in the early 1950's to persuade everybody to change over to cheap oil in the home? There was the advertisement for 'Mrs 1970' which carried it on effectively and very successfully, promising implicitly cheap oil forever and a day - for ten years. I can remember vividly a government White Paper of 1967, only twelve years ago, which said that the government considers it right to base fuel policy on the expectation that regular supplies of oil at competitive prices will continue to be available. So said the government in all its wisdom and I may add, using a very effective mathematical model for the first time, it came to the startling conclusion in 1967 that there was no real possibility of a major shortage of oil or a major increase in price; therefore no policy should be based on that assumption.

It's as well to remind ourselves of the more adventurous utterances made by our masters, if only in order for you to examine the adventurous utterances made in 1979! However, I have little doubt that there will be a major swing back towards the use of coal in the world as a whole. There is no real alternative in the sort of time spans that we are talking about. More about the time spans in a moment. Now, if you are going to have that sort of swing back to the use of coal, then you certainly want to be fully aware of the effects of what you are doing. Too often in this country, we find ourselves making policy decisions without a careful analysis of the implications of those decisions. On the one hand there is what I call the total global approach. You dream up in your bath a notion that can be described in grandiose terms as a new global policy and you leap up crying 'Eureka' or some other scientific comment and you say 'That's what we must do' and you don't do your homework properly. That's one way of arriving at a 'scientific policy'.

Another way is to start with the minutiae and look at and examine one particular problem in detail. Unfortunately that very method of examining that problem minutely can turn it into a major rather than a minor one and again you can find yourself not seeing the wood for the trees. In that way such an approach can be equally non-productive in arriving at sensible, workable solutions achievable in a reasonable time span.

I believe that you ought to know and the world ought to know what the environmental consequences of using more coal will be and this sort of conference should be first of many,

Some Environmental Aspects of the Production of Solid Smokeless Fuel

and I hope it will not be the last one that will invite a non-engineer or a non-scientist to speak to it.

I want to talk about not only the problems of making, that is manufacturing solid smokeless fuel, but also about the problems of the manager in reconciling desirable social objectives which happen to be in conflict. It would be a very nice world if our individual social objectives, desirable in themselves, all fitted into a pattern that gave you the right sort of overall picture - but they don't. With most of the problems that cross my desk relating to the environment I am faced with directly conflicting social objectives and, whichever one I pursue, I can find very good reasons, or have been told some very good reasons, why it is harmful in the context of another social objective. The analysis and balancing of social objectives, the decision-making, is something that we all too often wish to refer to other people normally known as 'they'. 'They' are the people that won't let me do this. 'They' insist that I do it that way. 'They' want me to do this and very often when you analyse it, it's an excuse, an excuse for not facing up to the problem yourself and making up your own mind what your own answer to that problem is.

Well, so much for a rather lengthy preamble. To justify my having come here, I think I have to talk about that area rather than the area that you have been talking about in some detail so far. I am not equipped to enter into a detailed debate with you about the science of the engineering aspects of the problems that we are faced with, but I think I am equipped to make some comments on the practical problems. While I have strayed from my theme rather a lot, I do intend to refer to solid smokeless fuel. I start by recalling yesterday morning listening to Glyn England who made a splendid historical comment about what happened in 1285, I believe. I think I can go back further. It strikes me that the demand for having a fuel in your home that is smokeless is very old indeed. I have a mental picture of prehistoric man going back to the cave dragging his beast of prey behind him by the hair, finding his mate in the cave and the mate saying, as all mates will, 'You'll have to do something about that damned fire - it's been smoking me out all day', and I am quite sure that the introduction of charcoal was regarded as being a major innovation in very early times. Then, later on, the realisation that there were certain types of coal, particularly anthracite, that burn smokelessly once you've got them ignited was an incredible advance and we were in a situation where the aim to improve domestic contentment was a strong motivation.

The elimination of smoke wasn't always simply a matter of domestic bliss, however, and it is a matter of some historical interest that the development of the Welsh dry steam coal industry was really based on the fact that it enabled the navies of the world that could afford to buy it to get that much nearer to their enemies without being betrayed by tell-tale smoke over the horizon.

However, in our search for domestic happiness, we ran into a period that involved humping great lumps of coal onto the fire and this led to smog and fog of really horrifying dimensions. Dickensian London, and I use the words advisedly, must have been a horrific place in which to live. If I can remind people of the opening of Dickens's 'Bleak House' where, in the first paragraph, he refers to

> 'smoke lowering down from chimney pots making a soft black drizzle, with flakes of soot in it as big as full-grown snowflakes gone into mourning one might imagine for the death of the sun.
> 'Fog everywhere, fog up the river where it flows among the green aits and meadows; fog down the river where it rolls defiled among the tiers of shipping and the waterside pollution of a great (and dirty) City'

There you are - 1850s - Dickens's 'great (and dirty) City' of London. Dickens was very perceptive in all sorts of ways; it is a pity he is not read more these days. He realised that the effects on the environment were not only on the individual but also on buildings which in their turn affected the living environment of people. Anyhow, if I can be allowed a second quotation, this is from 'Hard Times', when he was describing a rather appropriately named fictional town called Coketown, which I believe was, in fact, Preston. He commented on Coketown (and remember he published this in 1854 or thereabouts):

> 'It was a Town of red brick, or a brick that would have been red if the smoke and ashes had allowed it but as matters stood, it was a town of a natural red and black like the painted face of a savage. It was a Town of machinery and tall chimneys out of which interminable serpents of smoke trailed themselves for ever and ever and never got uncoiled'

Now he realised, like lots of other people, that not all the pollution was caused by coal or smoke in the atmosphere. He went on to talk about 'the black canal' and 'the river that ran purple with ill-smelling dye' and he had a word or two for town planners, as he commented on Coketown that:

> 'It contained several large streets, all very like one another, and many small streets, all very like one another, inhabited by people equally like one another who all went in and out of the same houses to the same sound on the same pavements, to do the same work and to whom every day was the same as yesterday and tomorrow and every year the counterpart of the last and the next'

Now, that is bringing the environmental issue into a wider context but a crucial context. For when you talk about how you might improve one aspect of the environment you must think about the environment as a whole and that means not just clean air, it means how people live, whether they have a happy existence, and any attempt narrowly to define the environment, in my view, whether related to the burning of coal or anything else, is doomed to overall failure because it will not get public acceptance.

It is surprising that given that description of Dickensian London it was such a long time before there was a general public recognition that smog was far more than a nuisance but was, in fact, a killer and therefore needed to be tackled. The interim report of the 'Committee on Air Pollution', published in 1953, drew attention to the harmful effects of smog on people's health and, in turn, led to the Clean Air Act of 1956, with which we are all familiar.

I find the 1956 Act very interesting, again in relation to what Mr England said yesterday morning; he made the comment that we were so far fortunate in the UK in that we did not have legislation narrowly defining the permissible levels of this, that, and the other and what you could do and what you couldn't do, and the Clean Air Act, which most people would agree has been a striking success, made no attempt at narrow definitions of that sort. It empowered the Minister to create smoke control areas in which it would be an offence for dark smoke to be emitted from a chimney of any building, but that it would be a defence to prove that the emission of smoke was not caused by the use of fuels other than an 'authorised fuel'. Therefore, if you could show you were using an authorised fuel, you were in the clear. Question: what is an authorised fuel? Well, I have enormous admiration for the British Civil Service and I think that the precise and scientific way in which they define an 'authorised' fuel takes a lot of beating. They define it as being 'a fuel declared by a Regulation of the Minister to be an authorised fuel'! Now I make the point really in two ways. If I were to read that out to an American audience, they would regard it as appalling; they want to see it in black and white in the law, in the regulations. If one were cynical, one would say that that was partly because they are a legalistic society dominated by lawyers and every time anything happens over there, the lawyers want to find some way of making money out of it. The fact remains, however, that no one said 'That's unworkable; it cannot work; it will be ineffective'. The question I am asking, ladies and gentlemen, is was it effective and the answer is, yes it was effective.

Well, what were the authorised fuels? They were by definition supposedly smokeless. The natural ones you are aware of: anthracite and dry steam coal; but the fuels produced naturally were, with the usual irony of things, those which were the hardest to produce, the most expensive you could get and above all mostly in limited supply. This is another example of 'Murphy's Law'. So you had to develop manufactured

Some Environmental Aspects of the Production of Solid Smokeless Fuel   175

smokeless fuel and this meant that you tried to build on what you already had because you did, in fact, already have some manufactured smokeless fuels, although the main one was produced as a by-product and was gas coke. This was of course merely the residue of coal used to produce town gas and it used to be a major domestic fuel, something which we have now forgotten.

As late as 1955, which is only a quarter of a century ago, the amount of coal used for gas making was 28 million tonnes and that led to quite a lot of gas coke. You remember gas coke - I remember it extremely well - it was one of the delights of my childhood when they delivered the coke to the school yard in this great mountain and then you could have an enormous amount of fun rushing up and down the coke pile, throwing pieces about and wondering why, in the end, you were filthy. Here I ought to make the point that gas coke didn't simply go off the market because of the introduction of natural gas. The use of gas coke was declining before then when you started having total carbonisation. Now, I have no wish to weary you immediately after lunch on your second day with all the formalities of how the coke was produced. The details are given in the Appendix.

How the domestic market has moved is extremely interesting and Figure 1 attempts to show how the market moved in the past 30 years in terms of millions of therms of domestic heating. Obviously, it takes no account of the improvement in efficiency in the use of therms between 1948 and 1978 but do not imagine that the total use domestically has not risen; that is not true. What you have here is the decline, or rather the elimination, of straightforward bituminous coal as a source of domestic heating. Oil stayed up. As town gas tailed off, it was replaced by natural gas. Electricity has grown. Solid smokeless fuel has gone through the odd bump here and there and now runs at around 12 million tonnes of coal, of which half is coal and half is manufactured fuel. When the production of gas coke died away, we in the coal industry started producing our own coke for sale to the domestic market, which was sold under the name of Sunbright. We call it a hard coke, and in effect,

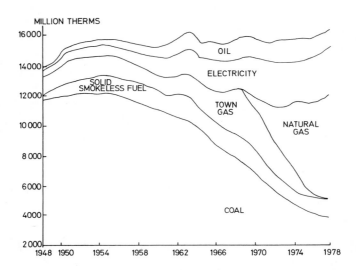

Figure 1 Contributions of various energy sources to domestic consumption

it was the undersized fractions from metallurgical coke manufacture of mainly conventional oven batteries. It was an attempt to try and meet what was a real gap in the market, and that's another theme I ought to come back to because you cannot ignore the fact that markets are there. You cannot ignore the fact that unless they are supplied, people physically go short of warmth, go short of hot water, and go short of what they regard as being an essential part of life. You cannot, therefore, take one extreme view and say, 'We do not like the adverse environmental effects of producing solid smokeless fuel; therefore, we will stop it or we'll make sure it is stopped tomorrow'. If you do that then you immediately run up against what I have referred to already, your first clash with social objectives. Whether people are right to want it or not, they do want it - they do, in fact, find that they have a use for it and they want it to continue to be produced.

Some Environmental Aspects of the Production of Solid Smokeless Fuel 177

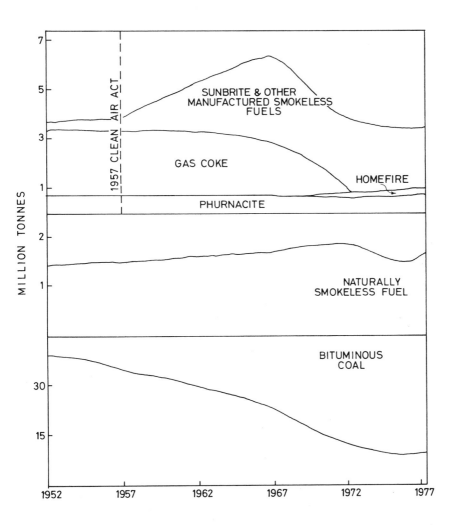

Figure 2 Disposals of solid fuel to the domestic market

The top part of Figure 2 shows how the markets for various products within the total solid smokeless fuel domestic market have moved. Note that immediately after the Clean Air Act of 1956 there was an increase in the total supply of solid smokeless fuels. We also see here the impact of the introduction and growth of oil and natural gas and the introduction of Phurnacite. One also has to refer to yet another premium fuel, Homefire, which came along later. This graph indicates that after the upsurge, which was caused simply because there was a market to be filled and people wanted solid smokeless fuel, the market fell away owing to the penetration of natural gas oil. But what happens as the Gas Board runs out of the ability to take on an increasing load? (I don't mean runs out of gas.) You have read in the Press that the Gas Board has already had to say that it is unable to accept major new industrial users; therefore, you have a situation where you cannot assume that the gas supply will be there to fill any gap on the domestic market. I need hardly mention oil and the price of oil to illustrate that you have a real market problem facing you.

I don't want to imply and I hope that I haven't implied that the National Coal Board are the only people who make solid smokeless fuel. Coalite and Rexco are very big and very active in the market.

I would now like to move on from coke production to briquetting processes. I have already mentioned Phurnacite. Now Phurnacite is a pitch-bound, roll-pressed ovoid briquette batch, carbonised at medium temperatures in Disticoke slot ovens, then cooled by a water quench. I simply give that description because it is apparent immediately that the way in which we produce Phurnacite is bound to lead to a number of local environmental problems. Here you have an accident of history. With hindsight, nobody would ever have put the Phurnacite process where it is - in a valley which suffers all the problems of temperature inversion, so that the dispersal through the atmosphere of the undoubted 'pollution' emanating from the plant is much inhibited by its sheer location. The location was decided in about 1941 or 1942 so it is not a new decision but it illustrates the problems you can inherit.

Incidentally, it illustrates the progress that we make in that there is no doubt whatsoever that if you were intending to start a new process on a brand new site now you would be taking into account precisely those factors that make the location of the Phurnacite plant so inappropriate. Having said that, the plant is there and it is providing work for over a thousand people and the first people who shout and raise their voices at any suggestion that you either run it down or move it somewhere else are the unions and the local authorities.

Homefire was an attempt to make in an environmentally more acceptable way a briquette really aimed at the open-fire market. These are self-bound char briquettes in which ground high volatile content coal is partially carbonised in a fluidised bed, heated at about 400 °C by combustion of some of the volatile matter with the fluidising air. The resulting low volatile char is compacted while still hot using an intermittent extrusion press to form short hexagonal cylindrical briquettes. These are cooled slowly for about an hour during which time the volatile matter content continues to decrease and the briquettes increase in mechanical strength and are then quenched in water. As far as the people living around it are concerned this is far more acceptable than the Phurnacite process. There is no doubt about that at all.

These are our current two briquetting processes but there are others. There is a German process called Ancit which we are currently considering as a replacement to the worn-out batteries at the Phurnacite plant. It involves blending low volatile non-coking inert coal with a third coking coal after the ground inert component has been heated in a hot gas stream to 600 °C.

I hope I am already giving the impression that when you produce solid smokeless fuel you may run into environmental problems in its manufacture because in all the various processes smokelessness and strength of the product are achieved by treating the coal to drive off more or less substantial proportions of volatile matter. I examine those environmental problems in two ways. Firstly, what are the effects on the people who actually work at the plant? (You had a fairly graphic illustration yesterday, I understand, of something which I can only say is news to me and does not tie in with any information I have, but perhaps I can come back to that later.) Secondly, what are the effects on the people who live around the plant?

I am extremely concerned that we should operate plants that are environmentally safe to both groups of people. You can quite properly say to me, 'What do you mean by safe, how safe?' and immediately you are faced with the balancing of social objectives. But for the moment I am describing this in terms of something in which there is no major risk to the health of the population, either working on the plant or working outside. In relation to coke ovens it is very interesting that experience and scientific information differ. Why they should differ is really a matter for you rather than me. What I find intriguing and worrying, in a number of ways, is the inconsistency of the evidence. For example, a detailed study has been carried out on the mortality of people who worked on coke ovens in the United States which seems to indicate (and on the basis of the report, certainly did indicate) a direct correlation between exposure to work on coke oven batteries and cancer. We immediately set in train a similar mortality study on our ovens in the UK, and we did this not by starting from the present but by going back over the records of all the people who had worked with us over I think the past 10 to 15 years, and died in the past 10 years, and analysed the data to see whether or not our findings were the same. If they were, then that would be a matter of major concern on which we would have to take direct action. But in fact our findings were diametrically opposed to those of the American study. There are a number of reasons why that should be so, because there are a number of variables which were not analysed, such as the type of people who work there, their life styles, their habits, etc. - the sorts of variables of which we are all fully aware but which are extremely difficult to quantify. But the stark fact indicated by the mortality study was that the immediate impression given by the American study was not transferable to UK experience, and this is an important point.

Now, it takes me on to my theme of the role of the scientist and engineer in these areas. I believe that in producing solid smokeless fuel we are doing two apparently contradictory things: we are making environmental conditions better in certain parts of the country, but making them worse in other parts.

Figure 3 shows where the solid smokeless fuel plants are located: most in the Midlands and Yorkshire, one in Wales, two in the North East, and one in Scotland. Now, these are the areas where you produce, but the bulk of the demand for manufactured solid smokeless fuel (see Figure 4) is something like 0.5 million tonnes in the South of England, 280,000 tonnes in the North, and a relatively small demand in Wales and the West of England. So, on the face of it, the people who say that if you produce in a given area the inhabitants have to suffer for the benefit of other people have an argument. The question is: are the inhabitants suffering and by how much? This is where, ladies and gentlemen, you come in. We endeavour to operate our plants in a way which minimises both the hazards internally and the effect on the environment externally. If you were to ask whether we could do more, the answer is, if we had unlimited sums of money at our disposal, I would hesitate to say that we could not do more. But the amounts of money that are already spent are by no means inconsiderable. If I can give an example: At Phurnacite, where we know we cannot make a major impact in wiping out the adverse effects on the local people, we are currently spending £4.5 million on measures which give no return whatsoever other than to make the environment better for the people living around. These are not safety measures, and they have nothing to do with the safety of the people on the plant, but they do make conditions a little more tolerable. Here we are straight into a dilemma. People say to me 'Why don't you shut it? How on earth can you justify operating a plant where by your own admission you are making a detrimental impact on the environment?' To which my answer is, 'What would happen if we did shut it? Nobody is arguing that

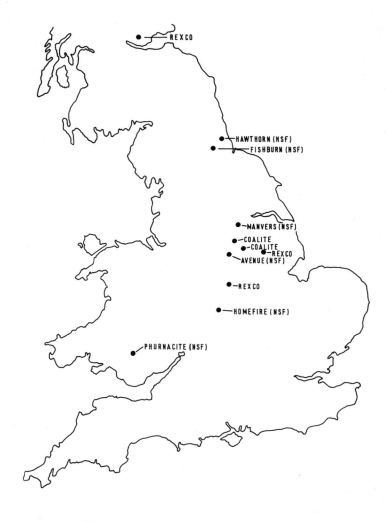

Figure 3 Locations of solid smokeless fuel plants

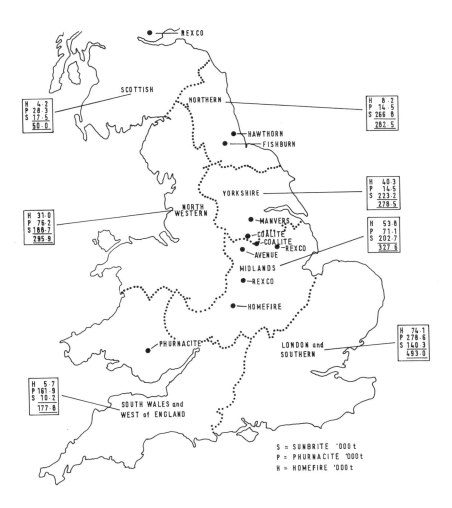

Figure 4 Use of manufactured solid smokeless fuel

it's a danger, nobody is saying that the health of the people on the plant is in danger or that of the people round about, and if we were to close it down 20% of all available manufactured solid smokeless fuel would disappear tomorrow. What would that mean in towns where the air has been greatly improved by the Clean Air Act?' I must say that I am a great supporter of the Clean Air Act and I am most anxious that we do nothing in this country that will detract from improvements in that area. I am saying, however, that there is no simple solution in relation to the actual plants themselves.

I would like to leave you with two or three thoughts. I, like most industrialists, am most anxious to know the true facts about the situation. I am most anxious to know whether or not we are genuinely endangering people. When I say that I mean I need to have accurate information about it. I do not want information that is biased and I do not want information that is really taking a very narrow view of the problem. Let me give you an example. I know of a smokeless fuel plant where there was a great complaint about the effect on the local inhabitants of the sulphur coming out of that plant. We had a strike and while the plant was shut down the level of sulphur in the town did not fall: it went up. Now I'm not saying that means that we are making a positive contribution, but I am saying that you have got to be very sure about the accuracy of your facts, particularly when you will be encouraging people to take decisions which can have effects that are not acceptable on other parts of the environment. We want facts and we don't want opinions. We want them carefully correlated and I am always wary of correlations. (I ought not to be wary of correlations, as it is the only area I know anything about, but I am wary of correlations.) For example, at the moment we have an economic policy that is based on getting down the amount of money in the economy on the grounds that that leads to inflation, and it is interesting to note that the best correlation is not between the amount of money in the economy and the rate of inflation, but between the outbreak of dysentery in Scotland and the rate of inflation! Now the figures fit almost perfectly but only an idiot would say that there is any direct relevance. I suggest that it would be equally wrong to assume that because you think that something must be damaging this or causing that effect and if you get a correlation that seems to justify that conclusion, you assume that it is right. If we had assumed, for example, that the American experience on the mortality of cokeworkers was right, we would have taken decisions that would have been totally wrong. We would have started doing things which in my view would not have been right and would have undoubtedly ended up with loss of employment and loss of activity. We have to spend money, and spend it wisely, particularly in a day and age when the words 'public expenditure' and 'social benefit' are really becoming dirty words. You hardly dare mention them, for 'public expenditure' is something that you shun and 'social expenditure' has to be kept down. Therefore, I'm not arguing that you shouldn't spend money on environmental areas; in fact, I spend a great deal of money on them. I am saying that you must spend it wisely, that you must spend it in areas where you know that you have a real problem, and can show that you have - then you spend a lot of money and you cure the problem, and if you can't cure it then you have to cease the activity. But we cannot afford, at a time when the world is hell-bent on galloping towards a major recession, to take the attitude that there is a sort of good fairy around the place with a plentiful supply of wishes that will enable you to do all that you would like to do the day after tomorrow. That is not the real world, and like so many other things, those who cry 'wolf' too often don't get the end result that they desire. I am really making a plea that industrialists and people involved in the environment work together to get the right answers, because we have shown already the major improvements that we can make to living conditions in this country by what I regard as a civilised approach. I am also pleading that we don't pursue our own particular hobby horse in a narrow area at the expense of the wider picture. I believe that the whole is greater than the sum of the parts, and that you do not get right answers simply by getting each individual part right; you have to take a total view. Now, I realise that this is controversial. I think, however, that we require a united effort to analyse environmental problems. I spend half of my time arguing with industrialists that we need to spend more time and more money on doing something for the environment. What I am anxious to do now is to make sure that we do not find ourselves oversold, and in that the contribution of a conference such as this is, I think, invaluable. It enables people to point to areas of concern. It enables people like myself, the scientific illiterates, to be able to see where there is concern and to see how we can measure it. But if you are going to be influential, then be sure of your facts to demonstrate an opinion. It may be a right opinion, it may be a correct opinion but, scientists and engineers, we want the scientists to analyse the problem and suggest solutions; we want the engineers to find cheaper and better ways of actually implementing those solutions.

You have heard, in the course of this conference, of ways in which burning coal can be made more effective and less environmentally worrisome. I believe that we are in a situation where there is a market for the manufacture of solid smokeless fuel. If we can find ways out of that, if we can get to a situation where they are not required, fine. If we work to it smoothly, fine. But I do not believe that we are anywhere near that yet, although there are developments on the way: we are developing methods of burning coal smokelessly. The point I am making is that you cannot make arbitrary decisions and chop things off without enormous effects on the wider environment. It is essential to protect the health of communities; it is essential to protect the health of the work people. It is not equally essential to destroy those communities, and you can destroy communities by taking well-intentioned measures in the short term before you have sorted out long-term problems, and my plea, ladies and gentlemen, would be, let's work together on those problems. Let's accept that there are problems there and let us realise that the resources that we use in this area we cannot use in other areas, and, in a time of limited resources, I don't think that we can ignore that fact. Thank you.

Appendix UK Production - Types and Processes

Types: Coke; Briquettes; Natural

Coke Processes

Gas Coke. Residue from coal used to produce town gas. Used to be a major domestic fuel: coal use for gas peaked at 28 mt y^{-1} in 1955 but has been virtually nil since 1975. Medium-temperature carbonisation (at 700-900 °C) in retorts was commonly used and resulted in 55-60% solid yield. This coke had 2-8% volatile matter and fairly high reactivity. (NB: Total gasification of coal, leaving no coke residue, began to displace medium-temperature carbonisation for gas making before coal itself was displaced as feedstock.)

Domestic Coke ('Hard' coke). The undersize fractions (doubles for open grates and openable appliances, smaller for closed appliances) from metallurgical coke manufacture. Made in conventional slot oven batteries. Sufficient reactivity for domestic use is a problem.

Sunbrite ('Hard' coke). Made in slot ovens with coal and operating conditions chosen specifically to produce a high-grade domestic coke.

Coalite. Made in vertical tube retorts. As in slot ovens, operation is batchwise and heating is indirect (by burning gas in tubes alternating with the retort tubes).

Rexco. This is originally a batchwise process, using vertical cylindrical retorts, with direct heating carried out intermittently. Hot gas, from a combustion chamber burning gas distilled from the coal, is passed downwards through the charge of coal; when the top 'half' is carbonised, combustion is stopped and continued circulation of gas, now unburnt and cold, is used to transfer heat downwards to carbonise the lower part of the charge. Cold gas circulation is continued to cool the coke to 100 °C before discharge. In a retort of later design continuous operation is achieved by allowing the charge to move downwards, passing through a gas-heated hot zone, then through a gas-cooled cold zone.

In general, medium-temperature carbonisation is used to produce coke-type solid smokeless fuels. Shorter duration coking has been tried in the search for high reactivity but there are mechanical strength and pollution problems on discharging the coke from the oven. Warmco, in which sodium carbonate was added to enhance reactivity was abandoned for reasons of pollution.

<u>Wood Charcoal</u>. This is now limited in the UK to a small amount of barbecue fuel.

Briquetting Processes

<u>Phurnacite</u>. Pitch-bound roll-pressed ovoid briquettes are batch carbonised at medium temperatures in 'Disticoke' slot ovens (which have an inclined floor to allow gravity discharge), then cooled by water quench.

(Multiheat was a similar briquetting process but used low-temperature carbonisation, carried out continuously in a trough-shaped fluidised sand bed. It was abandoned on account of non-consistent quality. However, the process is currently operated in France where a metal conveyor is used for the carbonisation.)

<u>Homefire</u>. Self-bound char briquettes. Ground high volatile content coal is partially carbonised in a fluidised bed, heated at about 400 °C by combustion of some of the volatile matter with the fluidising air. The resulting low-volatile char is compacted while still hot, using an intermittent extrusion press to form short hexagonal cylindrical briquettes. These are cooled slowly for about an hour, during which volatile matter content continues to decrease and the briquettes increase in mechanical strength and are then quenched in water. (Roomheat, a similar process save that the briquettes were formed to a pin-cushion shape by a roll press, was abandoned on account of failure to produce correctly formed briquettes of adequate mechanical strength consistently.)

Ancit. This is a German process currently under consideration for use in the UK. The two feedstock components, about 2/3 low-volatile, non-coking 'inert' coal and 1/3 coking coal, are blended after the ground inert component has been heated in a hot gas stream to ca. 600 °C. The blend is thoroughly mixed, thus heating and softening the coking coal, which acts as binder when the hot mixture is formed into ovoid briquettes by a roll press fed directly from the mixer. The briquettes gain strength and lose volatile matter during period of 'tempering' in a thermally insulated bunker and are then cooled by a water quench.

(Maxiheat, a bitumen-bound roll-pressed briquetting process, has been abandoned because the product is too smokey.)

Natural Smokeless Fuel

Anthracite. All the anthracites have low enough volatile matter content for them to be classed as smokeless fuels in the UK.

Dry Steam Coal. The less volatile of the dry steam coals also meet the UK definition of 'smokeless'.

Both categories are mined (deep and opencast) and prepared in conventional ways.

Discussion on Session IV and General Discussion

Chairman: Well, thank you very much indeed. The thought crossed my mind when you were speaking just now that the decision on these sorts of problems doesn't rest with the people here. In your area, it is going to rest with the housewife and I'm wondering how she is going to decide between you and Mr Dean and Dr Goalby as to how the local domestic unit is heated. Are you now going to start marketing smokeless fuel in competition with synthetic natural gas? This is going to be very interesting. I don't know that environmental factors are an important criterion in the housewife.

D J Davison: At the moment I am astonished to find that the market for our products (and remember that this is a market built up after a number of years of very successful and very efficient promotion of very cheap natural gas) has remained high and buoyant. We had, in fact, been assuming that it would run down and we had been planning for this but the outcome has been different. Without any exaggeration I could sell another 25% more Phurnacite if I had it, and currently I have no product which is in under-demand. Now this is at a time when we are already in competition. So, I am quite happy that the market will decide that one.

Chairman: Now, may I throw the meeting open to miscellaneous questions? I know there are one or two people with new questions they would like to ask, and now please, it is open to anyone.

R G T Lane (Sir Alexander Gibb and Partners): We have had an excellent series of papers and they have given us many alternatives for the use of quite a vast quantity of coal. Now these papers have brought out one or two points of apparent difference. For example, Mr England referred to the necessity for quite a big transportation of coal. Dr Gibson referred alternatively to the use of coal at the pithead, and therefore the transportation subsequently of less noxious materials, electricity, gas, or oil. Now, another point in regard to the use of coal at the pithead - could not the waste material go back in the ground where it came from? And another point, we have today modern methods of excavation which would allow us to make good caverns underground; at least, the tankage could go underground, probably the greater part of the manufacturing processes could go underground. Now, the point I have made here is that these papers have brought out alternatives and the question I want to ask is this: Mr Chairman, is there an international organisation or a national body in the UK which has these alternatives under study so that in due course a proper presentation of the alternatives can be made and a good decision made?

D J Davison: Could I comment about the putting of mining waste underground? A hoary old one and people naturally wonder why we don't do it. There are two points, which I think when I describe them are fairly obvious. When you are mining you are getting access to the coal underground and you are building your shafts for man-riding or for coal coming up and material going down. You have the problem of the sheer logistics of moving the material back down again, which do enormously add to the costs of construction of the mine. With a new mine there is a second problem, which you will be aware of, and that is the bulk factor when you take material from underground. About 15 years ago, the Board carried out, for a number of years, a pilot operation for pulverising the waste material and then taking it back underground and blasting it in under air pressure. In the way that they were doing it, this proved to be not at all successful. Were back stowage to be used it would add enormously to the cost of mining and frankly, if you had that sort of money available at the National Exchequer, I would far rather move the waste and reclaim the Wash, for example, than put it down underground again.

Your other question is related to a wider issue than coal. You will be aware that there is currently sitting the Commission on Energy and Environment which is looking at coal, and they have been examining these very areas, including putting the waste back underground. That I do not believe is an economically practicable possibility. The other possibilities I cannot comment on.

D J Davison: Can I make two comments which I didn't make because of shortage of time? I am sometimes asked if there is a shortage of solid smokeless fuel, why don't we import it? Now, I have no objection to importing solid smokeless fuel but there is a genuine issue here. I think if we are not careful, we will find ourselves avoiding our own environmental problems by exporting them; in other words, if you are not careful, you are in a situation where you say that we are not prepared to do the dirty job over here but it is alright for the X or the Y and the French or the Germans to do it. I think on the whole that when one is speaking about environmental problems in relation to coal one has to go outside the national power reserve.

You had a comment yesterday - the usual one about the CEGB causing acid rain in Scandinavia. You cannot, in my view, examine the coal issue on a purely narrow national basis. What you have to do is to be aware of what other countries are doing and in being aware of that I think you have to learn from their mistakes in the same way that I think you ought to learn from their successes.

E S Rubin (Cavendish Laboratory, Cambridge): I am not sure to whom specifically I should direct this question, but perhaps one of the panelists can respond. One of the things that seems to be missing from the groups of papers at the Conference - and perhaps doesn't exist yet - is a more comprehensive national picture of the environmental impact of using more coal. Most of the environmental impact discussions have dealt with individual situations - a particular power plant or a particular type of technology. On the other hand we have heard projections based on national energy plans of aggregate levels of increased coal use, some of which will be in power generation, but a great deal of which, according to Dr Gibson, will occur in the industrial sector. I am curious to know if there are studies of - or if people are concerned about - the aggregate environmental impacts of such energy scenarios. For example, where will the increased coal use be occurring? What levels of pollutant emissions will result? What are the local and national consequences of emissions of air pollutants, water pollutants, and solid wastes? Most importantly, will the current approaches to environmental control that have been characterised as having worked quite well to date still, in fact, be satisfactory if we envisage substantial increases in the use of coal in the future?

S T McQuillin (Department of the Environment - Joint Secretary of the Commission on Energy and the Environment): Chairman, I might just say a brief word in answer to that about the Commission on Energy and Environment's Coal Study. I am one of the secretaries of the Commission which is chaired by Lord Flowers. The terms of reference of the coal study are 'to examine the longer term environmental implications of future coal production, supply and use in the United Kingdom looking to the period around and beyond the end of the century, including likely new technologies and conversion to other fuels and raw materials'. So we are trying to look in a study which is going to take about - well, it will certainly take until the end of next year - of all aspects of the production and supply and use of coal in the United Kingdom.

That includes digging up the coal, and as Mr Davison mentioned we have been looking at problems like back stowage of waste and so forth, transporting it and storing it. One of the problems, certainly if a lot of coal is going to be used in industry, is that it has to be got into cities and has to be stored on land which in the old days might have been used for storing coal but is now taken up for other things. It can

be burnt and used, as we've heard, in a whole range of
different ways. I don't know that any of the speakers would
really think that we are going to construct 30 gasification
plants as well as goodness knows how many liquefaction plants,
goodness knows how many coal-fired power stations; it's a
question of studying the various options. Assessment of the
different patterns of environmental stress that could arise is
not at all easy and I am not sure whether we would be able to
produce that many answers to all the questions that arise.
It's easy enough to make lists of amounts of SO_2 produced and
even to model the effect on air quality - rather more
difficult to say what that means for my lungs - but it's at
least possible to make lists of environmental impacts. As
Mr Davison and others have pointed out, that doesn't
necessarily get you all that far. What one wants to know is
what the balance of advantages and disadvantages is, not just
of the impacts of that kind on the environment from different
patterns of development but taking account also of the
implications for people of having different amounts and
patterns of energy supply available to them. What would be
the disadvantage of not having extra coal to burn compared
with the disadvantage of having a lot of coal and producing a
lot of air pollution? All these questions are under
consideration by the Commission - as I say, chaired by Lord
Flowers, with members from a number of other bodies such as
the Royal Commission on Environmental Pollution, and from the
major energy industries. Dr Gibson is associated with the
Study from the Coal Board. We have taken evidence from a
wide range of people, and the Coal Board in particular has
prepared an enormous amount of documentation on some of the
subjects, indeed on some that we have been considering in the
last two days. I don't promise any easy answers, but at
least if people ask who is studying it, well we are studying
it.

Summing-up and Closing Address

Introduction

F. A. Robinson

It gives me great pleasure to welcome Sir Hermann Bondi to this Conference and to wind up for us. As I explained yesterday, he is not going to sum up what has been said at this Conference because he has not been here. What he is going to do is to give us a picture of the environmental problems as they appear to the Top Policy Committee. He is, as you know, Chief Scientist in the Department of Energy and holds a very vital position in the environmental and energy world in this country. He is also a Professor of Mathematics in Kings College, University of London and has filled a great many distinguished posts in Britain. He was, in fact, born in Vienna and educated at the Realgymnasium in Vienna and came to Britain in 1937 and was educated again at Trinity College, Cambridge! Since then, he has done many important things, based on his knowledge of mathematics, in the field of astrophysics and astronomy. He also carried out radar research during the war, and this country owes him a great debt of gratitude. I won't go on very much longer though I have a tremendous list of his achievements of which many people will know, but I think we are all looking forward now to having this final summing up of the environmental aspects of the situation in this country when oil is going to be replaced by other sources of energy at which, on this occasion, we are considering mostly, its replacement by coal.

Summing-up and Closing Address

Sir Hermann Bondi
CHIEF SCIENTIST, DEPARTMENT OF ENERGY, LONDON

Thank you very much, Chairman, for having absolved me from the impossible task of summing up what I have not attended though I have had a chance to glance at one or two of the papers and have certainly seen the titles. But I think it is worthwhile in a subject that you have discussed at considerable depth if I outline a little the perspective within which I see the question of increased coal burning in the UK. I think the only way one can look at this is with a global perspective and with a global energy perspective. The way I like to look at the energy situation of the world in, shall we say, the next fifty years, is this: the world is sharply divided now between the industrialised and the developing countries. The industrialised countries are the chief consumers of energy. They are very much more adaptable than developing countries, though I think this adaptability has not always been remembered. Certainly, we have to adapt; there will be plenty of groans and aches and noises and even tensions but nonetheless we must always remember that, in our situation in an industrialised country, all this is totally different from the developing world. I like to say, Chairman, that three years ago, in the summer of 1976, we had an almost unique drought. If we ask about the chief health effects of that drought, I think the blood pressure of a number of Water Board executives went up. But when a drought occurs in a remote part of Ethiopia or in the belt on the southern side of the Sahara, it isn't the odd blood pressure that goes up, it is thousands of people who die, it is hundreds of thousands who are up-rooted and perhaps can never return to their previous location and way of life. The developing countries are a much more sensitive part of the world than we are. If I adapt something that was said in a different connection elsewhere, 'if the industrialised countries get a slight cold, the developing ones catch pneumonia'. I believe this to be true, and one cannot talk about energy without remembering this fact. Again, we think of ourselves as very energy hungry, and in many a way this is true; in particular the materials we use - steel and aluminium - are very energy intensive. Transport, of which we use a great deal, in its various forms is very energy intensive and so on. But we must also remember that several of the great growth industries of our time are in no sense energy intensive. Education in the 1950s and 1960s had a boom such as few parts of the national economy have ever experienced and it is not a very energy-intensive industry, though it does require facilities.

The whole field of electronics is one that has expanded enormously with a resulting reduction in energy consumption. You all recall that, not so long ago, all TV sets had thermionic valves producing a great deal of heat. The change-over to solid-state electronics has greatly reduced energy consumption. Word processors and computers will not make up for this reduction in energy usage. In many ways it is possible for industrialised countries to grow and to improve the quality of life without having a proportionate increase or anything like a proportionate increase in energy consumption. For the developing world, this is not possible. Economic progress in the developing world is a great need for all of us and is not possible without substantial increase in energy consumption over the current very low levels. I don't say they have to follow the route of Victorian engineering with its tremendous reliance on power but, whichever route they go, they must increase from the present extremely low levels of energy use to markedly greater ones.

What I am trying to drive at is that whatever we in the industrialised world do, whatever we decide, will have a very much larger impact in the poorer countries than here. In particular, in the next thirty or forty years when oil supply will not keep in line with demand except through higher and higher prices, the more oil we consume, the less will the

developing countries be able to buy, and they will suffer
accordingly. Only a couple of days ago, I heard of the sort
of interesting and disturbing consequence of the rise in the
price of oil. In areas of New Guinea where the traditional
fuel has been firewood the population expanded and oil became
the fuel. Oil cannot now be afforded and people go out,
whether it is allowed or not, and cut down trees. In that
tropical climate, this is rapidly creating new deserts
because, as you know, without the protection of the trees,
the soil is simply washed away. It isn't a question of
tropical forests being replaced by agriculture, it is a
question of their being replaced by deserts. This is the
spectre that should haunt us in everything we decide to do.

This is neither the place nor the time to enter the
nuclear debate. All I want to say is that whenever nuclear
power in an industrialised country is discussed, it should be
remembered that since not building nuclear power stations
means more use of oil, the bill for this will certainly be
paid in the developing world and will be very painful, for
there the resulting lower availability and increased price of
oil will cause real suffering.

None of us can imagine a future for this or any other
industrialised country in which coal will not play an
immensely important and growing part. In a country as
fortunate as this that has a lot of coal, the worries of our
friends and partners in Western Europe who, in future, as to
some extent they already have today, will have to import
their coal, often from far away, should be remembered. The
future of the coal trade with all it implies for politics and
international stress and strain is a very important topic. A
coal-rich country like this should not forget that not all
industrialised countries are in as good a position as we are
and that many of the developing countries will also have to
burn a great deal more coal, even if they have to import it.

Now, we can go a little further in this. I know that a
great deal was discussed here of turning coal into secondary
fuels, fuels to which our engineering is much better adapted

Summing-up and Closing Address 199

whether they be liquid fuels for transport, smokeless fuels for burning in domestic fires, or synthetic natural gas for a clean fuel elsewhere. What is always the case is that the secondary fuel contains fewer calories than that from which it was made. In some circumstances, the loss is rather small; in others it can be pretty large. So by being more refined and by being more complex as we are bound to be in a complex industrialised country, by using more electricity, substitute natural gas and so on, we will always be increasing the use of primary fuel over direct burning. On the other hand, it must also be remembered, as I know was discussed here, that in a big plant of any kind, emission controls while still expensive are nothing like as prohibitive as they are in very small plants. So one comes around to the view that we have the knowledge and the technology to contain most of the inevitable products of coal burning - though not quite all as I'll discuss in a minute. But because the way one sees future coal burning is not by going back to 1922 but by doing things differently, the peculiar disadvantages that were suffered then need not be envisaged for the future. Of course, there are always problems. Any extractive industry has environmental problems. In a crowded and very much beloved country like this, these problems are a little more serious perhaps than in Queensland or Eastern Colorado.

I am not convinced that on the side of controlling this environmental impact of winning rather than of burning coal (which is your topic) we can say we have yet all the technology we would like to have for controlling environmental impact, but with coal burning, my feeling is that it all exists. It is a question of price and it is a question of price in two ways. First, there is the capital investment needed for cleaning up to any desired degree what you produce. Secondly the fact that thereby you are bound to diminish the amount of useful energy you get out per ton burnt - even though in many cases, this is only of modest significance. I am sure you have discussed these points at considerable length.

Two or three matters, I think, are there to which I want to refer particularly. One is the extraordinary differential attitude of society to different kinds of risk. The risk of the motor car is large, but nobody thinks about it when he gets into his car. The problem of minutely adding to the carcinogenic properties of the atmosphere is liable to cause people very much more concern. I don't want to say it is stupid; I just want to say that we must never believe that talking to people about expectation values will resolve all their problems and make them say 'that's marvellous, that doesn't bother me very much'. People do have differential feelings about different ways of being killed. We must make the work place utterly safe and reduce the hours of work so that people can go hang-gliding. This is the mood of the world in which we live. It is no use our trying to shrug it off and say it is stupid. We may try to change tastes a little but it is very difficult. As regards carcinogens (about which particularly the United States feels extreme anxiety) it should be remembered that oxygen is carcinogenic and that coal burning reduces the amount of oxygen in the air. We must be aware in the area of coal utilisation, no less than in the nuclear field, that just working out for our own satisfaction that a particular risk is negligible will not necessarily convince the opposition. We must take attitudes and outlooks very seriously indeed and think very hard about them. It is no use to adopt a 'bull at the gate' method.

We know that SO_2 has been a particular worry and is, to some extent, a continuing one. There are measures for controlling it, some of which fit into what we want to do anyway, such as fluidised bed combustion, whereas others require a very expensive outlay for what may be an unimportant minor improvement in the quality of the air and the acidity of the rain. Possibly one doesn't even know whether it would have any such effect. But to have an extreme attitude to this kind of thing, and to say that SO_2 is a bad thing and therefore cost what it may we must get rid of it, I don't think that is a sensible attitude. We must always remember that if we use more of another fuel rather than the most convenient and abundant one because we load it with what may be only marginally desirable emission controls, the bill, as I have said, will always be paid in the poorer countries of the world.

Summing-up and Closing Address

More difficult to control than SO_2 are the various nitrous oxides. They are not peculiar to coal burning - you can burn pure hydrogen in air and you get a particular lot of nitrous oxides which are very unpleasant. But, it is something to be watched and in a measured and reasonable way to be controlled if we can convince our fellow citizens to be measured in their demands.

Let me go on to what I regard to be the biggest enigma of them all, and that is CO_2. I know you have had a lecture here from Sir John Mason who is very much more knowledgeable on this than I am. However, I would like to stress two or three points; some of them I am sure he will have stressed, others perhaps not quite so much. First, it would be absurd to be alarmed by the CO_2 problem today. It is something we have certainly got to study; it is something that will take us quite some time to study since our ignorance of the CO_2 situation today is very great indeed. All we know is that CO_2 has been increasing these past hundred years and that the increase is approximately half of what our fossil fuel burning put into the air. Now that quite clearly means that fossil fuel burning is only one of the components in this equation, the others of which we understand very little, particularly perhaps the atmosphere/ocean interaction. There is the additional worry of the matter to which I already referred - the felling of tropical rain forests which lock up such enormously large amounts of carbon per square kilometre. It would be very hard for us in the industrialised world who felled the bulk of our forests hundreds of years ago in order to be able to produce enough food, to go and try to stop people in tropical countries following our line, but it is a great cause of worry; first, because tropical forests hold very much more carbon than temperate forests do; secondly, because in our case it has been replaced by intensive agriculture which itself locks up a not unreasonable amount of carbon. The development I fear is that in many of the tropical countries the period of agriculture on former forest land will only last a few years and then it will be a desert. This may not be true everywhere - there may be preventive measures but the desertification of the world whether through over-grazing with animals or through the felling of tropical forests is, I think, a major component of the worry.

Nobody knows how the carbon dioxide level varied in geological periods or what the effect on the climate was – nobody knows where the climate would be going if CO_2 was not increasing. If we are slipping into another ice age, perhaps we should burn everything, tables and chairs included, to warm the planet because only by increasing the CO_2 level can we improve our chances of keeping the climate tolerable.

But let us again remember that the atmosphere is an extremely sensitive machine, at best, metastable, that small changes in the atmosphere whose direct heating or cooling effects may be almost negligible could be enough to change the rainfall pattern. Over much of the tropical world, the rainfall pattern is not a question of how many umbrellas you sell, it is a question of life and death. The connection between how much we burn and the changing rainfall pattern may never be cleared up, but if a drought disaster occurs in the developing world we in the industrialised world may be accused of having been the cause. If we have a bad conscience because we are not quite sure that the accusation is wrong, we find ourselves with possibly much more awkward tensions to deal with than if the growing season in Canada lengthens or reduces.

I have talked very much of the global perspective; I think that in this global perspective this country has an important role: it is quite a significant component of the whole. What we do here affects others. How much do we want to sacrifice in order to avoid certain things which may or may not be harmful; we need to make sure that we spend our money and use our energy in the most sensible and sensitive way: that seems to me a major topic of research. A central feature of the whole business of modern technology, but particularly of energy technology, is its very long time scale. It is easy to forecast for ten years because any installation that hasn't started building now will certainly not be contributing in ten years time. If the world should (and I regard the probability as very low) but if the world should in twenty years time come to the conclusion that we must be a little careful of fossil fuel burning, if we can adjust to this conclusion in fifty years, we are remarkably clever. I regard one hundred years as the more likely period of adjustment. But the fact that there is a long time constant is not a reason for not doing the necessary research now in order to understand the problem. I have the feeling that our understanding of the weather machine, particularly perhaps its interactions with the ocean, is a priority subject for getting our energy policy sensible during the next century. The necessary increase of coal burning in this country will present some interesting questions for decisions that will not be quite easy to take. But otherwise their life would be very dull anyway.

Closing Remarks

Well, I think, ladies and gentlemen, that those very wise, all-embracing words make a very fitting end to the Conference. Sir Hermann, we have had a very good two days. As Dr Hislop said this morning, it isn't often you can get a meeting to mix up the scientists and engineers as we have done this time. It is only a few years since we began talking to one another and now we are talking quite regularly, and I am sure that it has been of tremendous benefit for each group to listen to what the other has to say and I hope that it is going to be of benefit to the country.

It is my duty and that of my colleagues to get on with the publication of the papers, and I hope that what we write will help to interest a wider audience in this very important topic of energy policy for this country. I am most grateful to you, Sir Hermann, for summing up in such a way what the energy problem is all about: it is not only what we have said here, but there are much bigger issues and I am grateful that you have brought in the international aspect and especially the problems of a policy for the third world. Thank you very much indeed.

That, I think ladies and gentlemen, concludes the Conference.